ATOMS
AND
ALCHEMY

ATOMS
AND
ALCHEMY

Chymistry
and the Experimental Origins
of the Scientific Revolution

William R. Newman

THE UNIVERSITY OF CHICAGO PRESS

CHICAGO AND LONDON

WILLIAM R. NEWMAN is the Ruth Halls Professor in the Department of History and Philosophy of Science at Indiana University. He is the author of *Gehennical Fire, Promethean Ambitions,* and, with Lawrence M. Principe, *Alchemy Tried in the Fire,* all published by the University of Chicago Press.

The University of Chicago Press, Chicago 60637
The University of Chicago Press, Ltd., London
© 2006 by The University of Chicago
All rights reserved. Published 2006
Printed in the United States of America

15 14 13 12 11 10 09 08 07 06   1 2 3 4 5

ISBN: 0-226-57696-5 (cloth)
ISBN: 0-226-57697-3 (paper)

Library of Congress Cataloging-in-Publication Data

Newman, William Royall.
   Atoms and alchemy : chymistry and the experimental origins of the scientific revolution / William R. Newman.
      p.   cm.
   Includes bibliographical references and index.
   ISBN 0-226-57696-5 (alk. paper)—ISBN 0-226-57697-3 (pbk. : alk. paper)
   1. Science—History.   2. Science—Philosophy.   3. Alchemy—History
4. Chemistry—History.   I. Title.
   Q125.N484   2006
   509—dc22

                                              2005030542

⊗ The paper used in this publication meets the minimum requirements of the American National Standard for Information Sciences—Permanence of Paper for Printed Library Materials, ANSI Z39.48-1992.

Contents

List of Illustrations | vii
Acknowledgments | ix
A Note on Terminology | xi

Introduction: The Problematic Position of Alchemy in the Scientific Revolution | 1

ONE The Mise en Scène before Sennert

1 The Medieval Tradition of Alchemical Corpuscular Theory | 23
2 Erastus and the Critique of Chymical Analysis | 45
3 Aristotelian Corpuscular Theory and Andreas Libavius | 66

TWO Daniel Sennert's Atomism and the Reform of Aristotelian Matter Theory

4 The Corpuscular Theory of Daniel Sennert and Its Sources | 85
5 The Interplay of Structure and Essence in Sennert's Corpuscular Theory | 126

THREE Robert Boyle's Matter Theory

6 Boyle, Sennert, and the Mechanical Philosophy | 157
7 Boyle's Use of Chymical Corpuscles and the Reduction to the Pristine State to Demonstrate the Mechanical Origin of Qualities | 190

A Concise Conclusion | 217
Bibliography | 227
Index | 241

Illustrations

Following page 82

1 Thin sheet of technical grade silver partially cut into small pieces

2 Silver pieces undergoing dissolution in nitric acid (aqua fortis)

3 Completely clear, blue solution of the technical silver in nitric acid

4 Solution of potassium carbonate (salt of tartar) being poured into the solution of silver and nitric acid to initiate precipitation

5 Precipitate consisting mostly of silver carbonate

6 Silver carbonate in a crucible after being filtered and washed

7 Hot silver that has been freshly reduced from the silver carbonate precipitate

8 Metallic silver in its crucible after cooling

Acknowledgments

The research for this book received generous funding from the Dibner Institute for the History of Science and Technology, the John Simon Guggenheim Foundation, the Institute for Advanced Study, the National Science Foundation (SES-9906126), and Indiana University. The original concept behind the book was born of the joint mentorship that I received as a graduate student from John E. Murdoch and Robert Halleux. My ideas received further stimulus from the conference on medieval and early modern corpuscular matter theories organized by Professor Murdoch and myself and held at the University of St. Andrews in the summer of 1996 (the conference was funded by Constance Blackwell and the Foundation for Intellectual History, and it resulted in the publication of Christoph Lüthy, John E. Murdoch, and William R. Newman, eds., *Late Medieval and Early Modern Corpuscular Matter Theories* [Leiden: E. J. Brill, 2001]). A subsequent seminar on Aristotle's *De generatione et corruptione* that I was able to attend in 1999–2000 under the direction of Professor Murdoch proved extremely valuable for the composition of the book as well.

I am particularly indebted to the detailed comments of three scholars who graciously gave their time to read and criticize a preliminary draft of the entire book—Peter Anstey, Michael Hunter, and Rose-Mary Sargent. In addition, I owe a debt to numerous other historians with whom I have discussed issues developed in the texts. Among them I include Domenico Bertoloni Meli, Michael Friedman, Noretta Koertge, Christoph Lüthy, Craig Martin, Margaret Osler, Lawrence M. Principe,

Alan J. Rocke, Alan Shapiro, Nancy Siraisi, Heinrich von Staden, Rega Wood, and the two students who offered valuable comments to various portions of the text used in a graduate seminar—Karin Findley and Cesare Pastorino. I also acknowledge the assistance of Cathrine Reck and Grant Goodrich in replicating Daniel Sennert's famous reduction of silver to its "pristine state." Thanks are also due to Catherine Rice and Maia Rigas, the acquisitions editor and copy editor responsible for *Atoms and Alchemy*. Their professionalism and helpfulness have greatly facilitated the publication and production of the book. Finally, I am grateful to my wife and family for tolerating my extended absence from our household in 1999–2000, when I was working feverishly at the Dibner Institute. Needless to say, the book takes issue with a number of received opinions, and the numerous scholars to whom I owe a debt cannot be held accountable for my final views.

A Note on Terminology

A number of terms used in this book require explanation, either because they are archaic and no longer generally used at all, or because the authors examined here used them in a way that is alien to the sense of a corresponding modern term, or because I have consciously chosen to use them in a nonstandard sense. The first term, "chymistry," requires little explanation, as Lawrence Principe and I have been using this archaic word for nearly a decade to refer to early modern alchemy-chemistry, a discipline that still viewed the transmutation of base metals into gold (*chrysopoeia*) as viable and yet contained much in addition that is identifiable to us moderns as chemistry.

More troubling is the term "atom," for this word and its cognates in the classical languages had different meanings even in premodern times. Since none of these meanings correspond to the highly complex atom of modern physics, the problem is further compounded. One of the principal themes of the present book is that chymists pioneered a particular use of the term "atom" and the idea behind it. Unlike the ancient atomists, such as Democritus and Lucretius, the early modern chymists, especially Daniel Sennert, used the term "atom" without the implication of absolute indivisibility, either mathematical (like a geometrical point) or physical (like the atoms of Lucretius). To Sennert, an "atom" was simply a chemically identifiable material unit that resisted further division by means of the operations of the laboratory known to him. Like many later chemists, Sennert assumed that there was no hard-and-fast distinction between laboratory processes and those in nature at large, so what is operationally indivisible in the one sphere should be so in the other sphere as well. Despite the fact that Sennert's usage prefigured more modern practice, numerous other chymists, both before and after Sennert, preferred words like "part" or "corpuscle" for such operationally indivisible particles, mainly in order to avoid association with the problematic mathematical implications of indivisible point atoms and to avoid being tainted by the atheistic implications of ancient atomism. Throughout this book I have respected the various authors' terminology, employing "corpuscle" and "particle" for writers like Geber and Robert Boyle when they eschew the word "atom" (although Boyle's position on this matter changes over time) and using "atom" where Sennert and others use the term. I often employ the expression "semipermanent" to indicate that a corpuscle or atom was believed to be chemically distinct and durable and yet to lack the absolute indivisibility of a Democritean atom. The context and accompanying notes will reveal these and further nuances to the reader.

A cluster of terms also stemming from ancient philosophy could prove confusing as well, namely, "mixture" and the accompanying term "composition" with its closely related cousin "compound." These terms derive from the Latin of the medieval scholastics (*mixtio, miscere* and *compositio, componere*), but their modern meanings have come to have a sense almost opposite to their medieval one. To Aristotle, "mixture" (sometimes qualified as "perfect mixture") meant a situation where the initial ingredients melded together to form a perfectly homogeneous state, whereas "composition" meant a state of juxtaposed particles, like wheat and barley shaken together in a jar. In the language of modern chemistry, to the contrary, "mixture" means a mechanical juxtaposition where no chemical bonding has occurred, while "compound" means the product of a chemical reaction, where a chemical bond holds the ingredients together. One thesis of this book is that the scholastic alchemists, unlike other scholastics who lacked experimental experience, came up with something approximating the modern idea of the chemical bond, which holds elemental particles together intimately and yet allows them to retain their identity and to be recaptured intact. The medieval alchemist Geber called the product of this type of bonding a "very strong composition" (*fortissima compositio*), distinguishing it both from the total homogeneity of an Aristotelian mixture and the mere juxtaposition of an ordinary Aristotelian composition.

One can also point to a handful of terms that I use in a way not strictly conforming to modern chemical practice. I speak of "reversible reactions," for example, to mean any series of simple inorganic reactions that can be reversed easily to regain the initial ingredients. This does not correspond to the sense in which modern chemists use the expression, namely, to mean a reaction that never comes to completion in a closed system. In short, my use of "reversible reaction" has nothing to do with the modern notion of chemical equilibrium but rather with the alchemists' rebuttal of the strict Aristotelian concept of "perfect mixture," according to which (at least in the eyes of the major scholastic schools of thought) there was no possibility of reversing the process that we now refer to as a chemical reaction. Another problem for those conversant with modern chemistry could lie in my use of the word "reduction." I use this term consistently in the way that alchemists used it, namely, to mean the "leading back" of a thing (in accordance with the Latin original of the word, *reducere*—"to lead back") to its original state after it has undergone some significant change. "Reduction" often had the more specific sense in alchemy of the isolation or extraction of a metal from a compound—often an ore. This old use of "reduction" lives on in modern mining and metallurgical literature, but the term as used in this book is obviously far more general than what is implied in the modern chemist's determination of oxidation states or the mental apparatus associated with redox reactions. A similar situation occurs with the alchemists' use of the word "sublime" (*sublimare* in Latin): the original word simply meant to "raise on high." Although the Latin alchemists commonly distinguished sublimation from distillation in that their sublimation normally began with a "dry" material (i.e., one that had not been dissolved in a solvent), they were not as fastidious as modern chemists, for the latter restrict sublimation to processes involving the direct passage from the solid to the vapor state, whereas alchemists frequently speak of liquids, such as mercury, subliming. As is my normal practice, I have retained the traditional alchemical usage. Other technical terms, especially those of medieval and early modern scholasticism, will be found throughout the book, but their explication will be easier to carry out in the broader context of the narrative.

The Problematic Position of Alchemy in the Scientific Revolution

At some time in the mid-1650s, Robert Boyle wrote an essay on the Holy Scriptures in which he discussed the compatibility of reason and faith. Boyle mentioned Francis Bacon and Daniel Sennert, among others, as champions of a thoroughly Christian and modern science.[1] An odd pairing, one might think. Over the three and a half centuries since the young Boyle made this association, Bacon has come to represent the founder of experimental science, while the name Sennert, even to most educated people, stands for nothing at all. Even Boyle himself soon shifted his allegiances. The appendix to his *Origine of Formes and Qualities*, published little more than a decade after he explicitly linked Sennert to Bacon, depicted the German "chymist," natural philosopher, and physician as a diehard representative of an outdated scholasticism.[2] Sennert's reputation has never fully recovered. What are we to make of this surprising situation? Had the juvenile Boyle simply made a mistake in linking Sennert to the new trends in science, which he quickly and rightly came to repudiate? Or did the more mature Boyle silently appropriate Sennert's natural philosophy while suppressing his earlier

1. Robert Boyle, *Essay of the holy Scriptures*, in *The Works of Robert Boyle*, ed. Michael Hunter and Edward B. Davis (London: Pickering and Chatto, 2000), vol. 13, p. 197. See Michael Hunter, "How Boyle Became a Scientist," *History of Science* 33 (1995): 59–103, especially pp. 76–77.

2. Boyle, *Free Considerations about Subordinate Formes* (appendix to *Forms and Qualities*), in Hunter and Davis, *Works*, vol. 5, pp. 449–452 and passim.

recognition of it? In fact, neither of these questions can receive a fully affirmative or fully negative answer. Although in recent years it has come to light that Boyle tacitly borrowed fundamental aspects of his experimental corpuscular theory from Sennert, he also significantly modified the German academic's ideas.[3] But the issues embodied in Boyle's use of Sennert point in a remarkably clear way to our own poor understanding of the role that chymistry or alchemy played in the Scientific Revolution more broadly.[4]

At a time when even the term "Scientific Revolution" has become a contentious topic among historians of science, it may seem either impetuous or otiose to raise the issue of chymistry's place in this putative historical period. The charge of rashness could derive from the fact that some prominent historians of early modern science now view the "Scientific Revolution" as a concept that is incompatible with a fully contextualized picture of the various sciences in the early modern period. Steven Shapin hints at this in the opening lines to his survey of the subject— "There was no such thing as the Scientific Revolution, and this is a book about it."[5] Whether one accepts the term "Scientific Revolution" or not is of little consequence for my narrative, however. What I hope to show is the pivotal role that alchemy played in the great disjunction between the common view of matter-theory before and after the mid-seventeenth

3. For the establishment of Boyle's debt to Sennert, see William R. Newman, "The Alchemical Sources of Robert Boyle's Corpuscular Philosophy," *Annals of Science* 53 (1996): 567–585. The topic is also developed at some length in William R. Newman and Lawrence M. Principe, *Alchemy Tried in the Fire: Starkey, Boyle, and the Fate of Helmontian Chymistry* (Chicago: University of Chicago Press, 2002), pp. 18–22. But it cannot be said that Boyle merely appropriated this material tout court, since he converted it to fit his own mechanical philosophy. Precisely how Boyle transformed Sennert's experimental corpuscular theory is one of the central themes of the present book.

4. I will use the terms "alchemy" and "chymistry" as synonyms throughout this book, in conformity with the linguistic convention most prevalent in the early modern period. At the same time, the reader should be aware that the field denominated by these two terms did not concern merely the transmutation of base metals into gold and silver (*chrysopoeia* and *argyropoeia*), but according to its practitioners also encompassed iatrochemistry (*chymiatria*) and chemical technology more broadly. See William R. Newman and Lawrence M. Principe, "Alchemy vs. Chemistry: The Etymological Origins of a Historiographic Mistake," in *Early Science and Medicine* 3 (1998): 32–65. See also Newman, *Gehennical Fire: The Lives of George Starkey, an American Alchemist in the Scientific Revolution*, rev. ed. (Chicago: University of Chicago Press, 2003), pp. ix–x, 84–91.

5. Steven Shapin, *The Scientific Revolution* (Chicago: University of Chicago Press, 1996), p. 1. For more recent discussions of this theme, see Mario Biagioli, "The Scientific Revolution Is Undead," in *Configurations* 6 (1998): 141–148; and Peter Dear, "The Mathematical Principles of Natural Philosophy: Toward a Heuristic Narrative for the Scientific Revolution," *Configurations* 6 (1998): 173–193.

century. The fact that such a rupture did take place is beyond dispute. Acceptance or rejection of the label "Scientific Revolution" is no more relevant than whether we continue to use other convenient historical place markers such as "the Middle Ages," "the Renaissance," and "the Enlightenment." All such terms imply some significant historical change that can act as a point of chronological division. Although I accept the division of history into identifiable segments, it does not follow that the segments are dissociated from their neighbors any more than a commitment to the reality of biological cells impedes a belief in the existence of organisms composed of those cells.

Several recent historical studies have already addressed alchemy in the early modern period. In addition to my own work and that of Lawrence Principe (some of which we have cowritten), one could point to two very current studies—Bruce Moran's *Distilling Knowledge* and Pamela Smith's *The Body of the Artisan*—which both consider the role of alchemy in the early modern world.[6] Yet neither of these works focuses on the role of matter theory, which I view as having undergone a sea change at the hands of medieval and early modern alchemists. As the present book illustrates, it was alchemy that provided corpuscular theorists with the experimental means to debunk scholastic theories of perfect mixture and to demonstrate the retrievability of material ingredients. A further monograph, Antonio Clericuzio's *Elements, Principles and Corpuscles*, does give a convenient introduction to early modern matter theory and its relationship to chymistry. As I show in a later chapter, however, Clericuzio's interpretation of the mechanical philosophy—key to any understanding of the role of chymistry in Robert Boyle's thought—is open to serious criticism on a number of fronts. I see Boyle's mechanical philosophy as having been indissolubly linked to his chymical researches, whereas Clericuzio's work erects an artificial dichotomy between Boyle's mechanism and his chymistry. In addition, like most who have written on the

6. Moran's *Distilling Knowledge,* a useful popularizing study, does point to the importance of alchemical corpuscular theories deriving from the medieval alchemist Geber, but Moran does not present this material against the backdrop of continuist theories of mixture or describe the empirical basis of alchemical corpuscular theory. See Bruce Moran, *Distilling Knowledge: Alchemy, Chemistry, and the Scientific Revolution* (Cambridge, MA: Harvard University Press, 2005), 36. Pamela Smith's *The Body of the Artisan,* on the other hand, is an attempt to revive and expand the theory of Edgar Zilsel, first promulgated in the 1930s and 1940s, that the social rise of artisans was a necessary precursor to the Scientific Revolution. Smith puts new emphasis on the fine arts and alchemy, however, which were largely neglected by Zilsel. See Pamela H. Smith, *The Body of the Artisan: Art and Experience in the Scientific Revolution* (Chicago: University of Chicago Press, 2004). Also see my review of Smith in *Chemical Heritage* 23 (2005): 44–45.

corpuscular matter theories of the seventeenth century, Clericuzio fails to give scholastic matter theory its due.[7] Without an understanding of the theoretical bases that underlay the centuries of scholastic reasoning before the seventeenth century, we can neither appreciate the constraints within which the first laboratory-oriented corpuscularians operated nor can we understand the attractions of their new approach.

In fact, few areas reveal the great divide that separates us from the mainstream medieval and Renaissance view of nature so effectively as the theory of matter and its operations. At the beginning of the seventeenth century, only a generation before the birth of Isaac Newton, atomism was not widely upheld in Europe. Indeed, precisely the opposite was accepted by most of the learned community as an article of faith. Material change was generally explained not by the association and dissociation of microscopic particles but rather by the imposition and removal of immaterial forms. The very possibility that matter could be composed of invisibly small corpuscles having a more or less permanent character was routinely denied.[8] More than this, it was commonly believed that the ingredients of "genuine mixtures"—many of which we would today call "chemical compounds"—were not capable of being retrieved from their combined state at all.[9] Instead, it was a widespread tenet among natural philosophers, theologians, and physicians that the immaterial forms accounting for the qualities in a given portion of matter would undergo a corruption upon their replacement

7. Antonio Clericuzio, *Elements, Principles and Corpuscles: A Study of Atomism and Chemistry in the Seventeenth Century* (Dordrecht: Kluwer, 2000).

8. For some telling examples of prominent natural philosophers in the sixteenth and seventeenth centuries who explicitly upheld the absolute homogeneity of mixtures (and therefore the nonatomic, impermanent character of their constituents), see Hans Kangro, *Joachim Jungius' Experimente und Gedanken zur Begründung der Chemie als Wissenschaft* (Wiesbaden: Franz Steiner, 1968), pp. 127–130.

9. Strictly speaking, the domain of Aristotelian mixtures was larger than that of modern chemical compounds, since the modern elements of the periodic table would themselves have been considered "mixtures" by the scholastics: a metal, for example, was seen as a mixture made up of the four elements, fire, air, water, and earth. One must also add that the scholastics believed their "true mixtures" to be totally homogeneous, unlike modern chemical compounds that are made up of atoms held together by a chemical bond. Nonetheless, although the isomorphism between the scholastic antithesis of "true mixture" versus mere juxtaposition and the contrast that modern chemistry draws between a chemical compound and a mechanical mixture is problematic, it is true that the majority scholastic opinion forbade the retrievability of the initial ingredients both from substances that modern chemistry views as elemental and from materials that modern chemistry sees as chemical compounds.

by a different form, which would render the recovery of preexisting ingredients impossible. Despite the simplistic accounts of Aristotelian elemental theory often given by modern scholars, it was believed by a multitude of sixteenth- and seventeenth-century scholastics—especially those influenced by Thomas Aquinas and his heirs—that the thermal decomposition of wood or of any other material did not separate or expose its original constituents in any way. The flame, smoke, moisture, and ash revealed by combustion could not be the original four elements that went into composing the wood, since those had been transmuted by the form of the wood when it was imposed upon the elements during mixture.[10] Regardless of these roadblocks in traditional matter-theory, however, by the time that Newton had reached his maturity it was possible to assume without serious argument that matter was composed of minute yet robust corpuscles and to uphold the reversibility of chemical combinations with similar ease. It is in fact impossible to imagine Newton's successes in optics or physics as a whole without the heuristic assumption that beneath the threshold of sense, matter—and even light—are composed of discrete and permanent particles rather than a single, mutable continuum. The case is even more compelling if we push our enquiry as far as nineteenth-century chemistry, the golden age of the discovery of new elements and of the laws governing their interactions. Can one even imagine the work of Dalton or Berzelius without the underlying belief that enduring material units with identifiable chemical characteristics were responsible for chemical composition?[11] Where then did the transmitters of that belief, such as Newton, or for that matter Robert Boyle, the popularizer of the term "mechanical philosophy," obtain their full conviction that the microworld is corpuscular in structure? As this

10. Pace Allen G. Debus, who unfortunately relies on Robert Boyle for the details of Aristotelian mixture theory. See Debus's otherwise illuminating article, "Fire Analysis and the Elements in the Sixteenth and Seventeenth Centuries," in Allen G. Debus, *Chemistry, Alchemy and the New Philosophy, 1500–1700* (London: Variorum Reprints, 1987), VII, pp. 127–147, especially pp. 130–131.

11. The belief in durable chemical entities that resisted laboratory analysis and that reacted with one another in integral multiples was a sine qua non of chemical atomism in the nineteenth century, even when chemists either refused to speculate further on the nature of substances at the microlevel or hoped for an underlying *protyle* that subsisted beneath the level of elemental differentiation. An excellent treatment of this subject may be found in Alan J. Rocke, *Chemical Atomism in the Nineteenth Century: From Dalton to Cannizzaro* (Columbus: Ohio State University Press, 1984); see especially Rocke's definition of chemical atomism (pp. 10–14), and his treatment of Dalton (pp. 21–47) and Berzelius (pp. 66–78).

book will show, a large part of the answer lies in the realm of medieval and early modern alchemy.[12]

THE PLACE OF ALCHEMY IN TRADITIONAL ACCOUNTS
OF THE SCIENTIFIC REVOLUTION

And yet a reading of the existing survey literature on the Scientific Revolution gives practically no inkling of the role that alchemy played in that transformation.[13] Older accounts of the mechanical philosophy, with a few important exceptions, routinely picture Boyle and his peers as drawing mainly on the self-styled opponents of Aristotle such as Gassendi and Descartes and on the ancient atomists for their corpuscular theory—certainly not on alchemists. The most influential proponent of this perspective has undoubtedly been Marie Boas Hall, whose "The Establishment of the Mechanical Philosophy" (1952) is still widely cited as a definitive study of early modern corpuscular theory. Boas Hall, however, erected an anachronistic dichotomy between early modern "chemistry" and "alchemy" that served to legitimize the former as the discipline of Robert Boyle while casting the latter in the role of obscurantist impediment to scientific progress. Similar views permeated the historiography of science in the 1950s and 1960s, and as we will see, they are present in the current literature as well, both as implicit assumptions and as overtly cited credos.[14] More generally, the broad picture that we have received of scientific change in the early modern period is one that emphasizes

12. Newman, "Alchemical Sources," and "Robert Boyle's Debt to Corpuscular Alchemy," in *Robert Boyle Reconsidered*, ed. Michael Hunter (Cambridge: Cambridge University Press, 1994), pp. 107–118.

13. Although there is a growing appreciation of early modern alchemy among historians of science and philosophy, I refer here to the genre of surveys of the Scientific Revolution, which continue to recapitulate the views of a previous generation. For a brief conspectus of the recent literature on chymistry, see the new foreword in the 2003 edition of Newman, *Gehennical Fire*.

14. Marie Boas [Hall], "The Establishment of the Mechanical Philosophy," *Osiris* 10 (1952): 412–541. But see also Robert Hugh Kargon, *Atomism in England from Hariot to Newton* (Oxford: Clarendon Press, 1966), passim. These authors must be excused as having been victims of the prevailing opinion of their times, but the unduly negative view of alchemy resurfaces in many more recent treatments as well, as in the otherwise impressive book by Peter Alexander, *Ideas, Qualities and Corpuscles* (Cambridge: Cambridge University Press, 1985), p. 9: "The natural philosophers attacked by Boyle were largely influenced by scholastic views or by certain alchemical ideas or both. Boyle was strongly influenced by early and recent atomism, which had never become generally accepted largely because it was regarded as atheistical, it was attacked by Aristotle, and was difficult to make consistent with well established Aristotelian views." Here we find the commonplace historiographical dichotomy between alchemists and scholastic Aristotelians on one side and atomists both ancient and modern on the other.

above all the discoveries made in the fields of astronomy and physics—be it in Copernicus's new heliocentric system, Galileo's treatment of motion, the mechanical philosophy of Descartes and Boyle, or of course the polymorphous work of Isaac Newton. Perhaps it is not surprising that in the age of the Manhattan Project and the subsequent attempt to outdo *Sputnik*, the mid-twentieth-century historians of science who were largely responsible for the professional formation of the field had little or no interest in alchemy. Yet this does not of itself account for the derision with which Herbert Butterfield, in his celebrated *Origins of Modern Science* (1949) would dismiss historians of alchemy as being "tinctured with the same type of lunacy they set out to describe."[15] Nor does it explain the fact that A. Rupert Hall, in his *The Scientific Revolution, 1500–1800* (1962) denied alchemy any status as a forerunner to chemistry and went so far as to describe it as "the greatest obstacle to the development of rational chemistry."[16] Even E. J. Dijksterhuis, whose important work *The Mechanization of the World Picture* (appearing in Dutch in 1950 and in English in 1961) gave intelligent descriptions of alchemical corpuscular theory and the medieval theory of mixture, could only see folly in Boyle's pursuit of the aurific art. As Dijksterhuis put it, "in [Boyle's] case too alchemy remained what it had always been: a mysterious trifling with impure substances, guided by mystical conceptions and hazy analogies, in which credulity played a considerable part."[17] The bitter denunciation

15. Herbert Butterfield, *The Origins of Modern Science, 1300–1800* (New York: Macmillan, 1951), p. 98.
16. A. Rupert Hall, *The Scientific Revolution, 1500–1800: The Formation of a Modern Scientific Attitude* (Boston: Beacon Press, 1962), p. 310. The first edition of Hall's book appeared in 1954.
17. E. J. Dijksterhuis, *The Mechanization of the World Picture*, trans. by C. Dikshoorn (Oxford: Oxford University Press, 1961); this book remains a model worthy of emulation in several respects. Dijksterhuis was aware of various medieval concepts of mixture as well as alchemical corpuscular theory (pp. 200–209), suspected the influence of Daniel Sennert on Boyle (pp. 282, 436, 439), and recognized Boyle's ongoing involvement with alchemy (p. 440). Nonetheless, in addition to his wrongheaded view of Boyle's chrysopoetic inclinations, he could still utter the following dismissive and ill-considered words about medieval alchemy (p. 160): "The whole subject, which originally had fitted in sufficiently well with general philosophical and scientific thought to be entitled to a certain amount of understanding and appreciation, thus descended more and more to the rank of a rather unedifying record of the degeneration of science. In order to study it seriously, one must decidedly either be interested in the pathology of thought or, with C. G. Jung, be able to relate the subject to the areas of present-day depth-psychology." For the inadequacy of Jungian psychology as a tool for analyzing alchemy, see William R. Newman, "'*Decknamen* or Pseudochemical Language'? Eirenaeus Philalethes and Carl Jung," *Revue d'histoire des sciences* 49 (1996): 159–188, and Lawrence M. Principe and William R. Newman, "Some Problems with the Historiography of Alchemy," in William R. Newman and Anthony Grafton, *Secrets of Nature: Astrology and Alchemy in Early Modern Europe* (Cambridge, MA: MIT Press, 2001), pp. 385–431.

of alchemy expressed by these authors reflects a long historiographical tradition whose roots, ultimately, find their origin in the rejection of the subject by eighteenth-century chemists and philosophes eager to detach themselves from a dark and "irrational" past.[18] A similar position was taken by Marie Boas Hall in her celebrated *Robert Boyle and Seventeenth-Century Chemistry* (1958). To Boas Hall, the mechanical philosophy of Boyle owed little or nothing to alchemy, embodying "a new chemistry" in which there "was incorporated a physicist's view of matter." For Boas Hall, alchemy was a "mystic science" opposed to the rational developments of the seventeenth century, and thinkers like Daniel Sennert had little input into the formation of its antithesis, the "new science." To quote Boas Hall, Sennert "contributed nothing new to the development of a mechanical philosophy based upon a theory of atoms." Indeed, he was "neither original, successful, nor, ultimately, influential."[19]

It is fair to say that the traditional view of alchemy represented by Butterfield, the Halls, and many other writers on early modern science has found a new home in the most recent surveys of the Scientific Revolution.[20] The failure to recognize a role for chymistry in the development of the mechanical philosophy has not been rectified in the most recent comprehensive treatments of the subject, despite the emphatic claims to originality made for newer historiographical approaches such as the sociology of scientific knowledge.[21] The well-known *Leviathan and the Air Pump* (1985) of Steven Shapin and Simon Schaffer explicitly views alchemy as a foil to the experimental science of Boyle and the

18. Principe and Newman, "Some Problems with the Historiography of Alchemy."

19. Marie Boas [Hall], "The Establishment of the Mechanical Philosophy," *Osiris* 10(1952): 412–541, see p. 428; Boas [Hall], *Robert Boyle and Seventeenth-Century Chemistry* (Cambridge: Cambridge University Press, 1958), p. 75; Boas [Hall], *The Scientific Renaissance, 1450–1630* (New York: Harper Torchbooks, 1962), pp. 166–167, 263, n. *.

20. One could also point to many another synthetic treatment of the Scientific Revolution that views alchemy as an irrational bit player in early modern science. Richard Westfall's *The Construction of Modern Science: Mechanisms and Mechanics* (Cambridge: Cambridge University Press, 1977), pp. 68–69, adopts the view that alchemy was essentially wedded to a vitalist or animist concept of nature and that the subject could be integrated with the mechanical philosophy only when "alchemy" gave way to "chemistry." This artificial dichotomy, seemingly based on the work of Hélène Metzger, has been discredited in Newman and Principe, "Alchemy versus Chemistry," pp. 33–38.

21. An exception to this tendency may be found in John Henry, *The Scientific Revolution and the Origins of Modern Science* (New York: St. Martin's Press, 1997), pp. 42–72. Henry does consider the corpuscular views of such authors in the alchemical tradition as Geber and Sennert, albeit in passing.

Royal Society.[22] Shapin's recent survey, *The Scientific Revolution* (1996), merely reinforces this point. Alchemy makes a brief appearance here among the "pseudosciences," whose interaction with the "proper sciences" such as chemistry was "intensely problematic." Shapin may be relating what he views as broad seventeenth-century categories, but if so, he is badly mistaken. In fact, the imposition of a meaningful distinction between alchemy and chemistry is highly anachronistic for most of the seventeenth century, and especially for Boyle, whose transmutational quest extended from his earliest laboratory training at the hands of the American chymist George Starkey up until his death in 1691.[23]

Shapin's imposition of modern categories onto seventeenth-century chymistry is particularly ironic in view of his own extensively argued case for a "contextualist" history of science that would avoid the anachronistic excesses of those historians who have focused on the internal development of their subject. One might expect that Shapin's oft-stated respect for historical context and actors' categories would have steered him away from employing the dated yet modern distinction between "pseudoscience" and the so-called "proper sciences." Yet a closer reading of his theoretical writings reveals a point of paramount importance that helps to explain this lapse—Shapin's method consists largely of adding sociological explanations to the preexisting history of ideas rather than subjecting the results of intellectual history to critical analysis. Indeed, Shapin himself admits this relation of dependency, pointing out that "contextualists" should rely upon the "empirical findings" of "intellectualist" history. Shapin advises that the work of intellectual historians provides "the necessary starting-points for historians who would put an additional set of contextual questions to the materials." Although Shapin's "contextualists" need not accept the models of causal agency employed by intellectual historians, "they must build upon intellectualists' empirical findings."[24] Both in theory and in practice, then, Shapin's sociology of scientific knowledge occupies a second-tier, even derivative,

---

22. This has been commented upon at some length by Lawrence M. Principe, *The Aspiring Adept: Robert Boyle and His Alchemical Quest* (Princeton: Princeton University Press, 1998), pp. 107–111.
23. Steven Shapin, *The Scientific Revolution* (Chicago: University of Chicago Press, 1996), p. 6. For the unsustainability of the alchemy-chemistry distinction in Boyle, see Newman and Principe, "Alchemy versus Chemistry," p. 58. For Boyle's career-long involvement in traditional chymistry, see Newman and Principe, *Alchemy Tried in the Fire*; and Principe, *The Aspiring Adept*.
24. Steven Shapin, "Social Uses of Science," in *The Ferment of Knowledge*, ed. G. S. Rousseau and Roy Porter (Cambridge: CUP, 1980), pp. 93–139; see p. 111.

relationship to the existing body of intellectual history. This is what Principe and Margaret Osler mean when they speak of his work as "old wine in new skins."[25]

But this situation poses an interesting dilemma. What does the sociologist of scientific knowledge do when the received opinion transmitted by the intellectual historians changes? What happens when a field like alchemy ceases to be viewed as a scientific pariah and comes to be seen by intellectual historians as having made a serious contribution to the core elements that constitute the traditional narrative of the Scientific Revolution? In such a situation, the Shapinesque sociologist must either confront the rude truth that he has provided a social explanation for a "fact" that had no real basis or he must reaffirm the correctness of an outdated historiography. To judge by Shapin's writings, it is the latter path that is the more appealing. As Rose-Mary Sargent has pointed out, Shapin's arguments "betray an assumption that the history of ideas is in some sense complete."[26]

This attitude, at least in regard to the Scientific Revolution, may be seen both in Shapin's work and in that of his closest defenders, such as the historian of early modern science Peter Dear. Dear, another recent entrant into the survey genre, takes an approach similar to Shapin's (and to the earlier survey writers) in his *Revolutionizing the Sciences* (2001). Like Boas Hall with her "mystic sciences," Dear views alchemy as essentially secretist and bound up with the spiritual perfection of the would-be adept, unlike the emerging science of chemistry.[27] This outdated view is only intensified when Dear considers Andreas Libavius, among the most important writers in the early modern genre of the chymical textbook. To Dear, Libavius is not an alchemist at all, but "an important chemist," since "he drew stark, explicit contrasts with the secretive labours of the closeted alchemist." Yet this position is not only anachronistic but outright wrong, in a way that further highlights the misleading character

25. Lawrence M. Principe, "Boyle's Alchemical Writings: Anonymity, Uncertainty, and Oblivion," presentation given at the annual History of Science Society conference, 1993; the fact that Shapin is merely putting "old wine in new bottles" has also received notice from Margaret J. Osler in the introduction to her *Rethinking the Scientific Revolution* (Cambridge: Cambridge University Press, 2000), pp. 3–22, especially pp. 19–20 and note 52.

26. Rose-Mary Sargent, *The Diffident Naturalist: Robert Boyle and the Philosophy of Experiment* (Chicago: University of Chicago Press, 1995), p. 6.

27. Margaret Osler also complains of Dear's "dismissive account of alchemy," and John Henry criticizes the book's general skewing towards physics and mathematics to the neglect of other fields. See their reviews of *Revolutionizing the Sciences* in *Annals of Science* 61 (2004): 134–136 and *British Journal for the History of Science* (June 2004): 199–200.

of the alchemy-chemistry dichotomy. In reality, Libavius was not only one of the chief spokesmen for *chrysopoeia* (transmutational alchemy) of the early seventeenth century, allying himself explicitly with the medieval alchemical writings attributed to Geber and Ramon Lull, but he was also an inveterate aficionado of the very secretism that Dear believes him to have decried.[28] Libavius spent countless hours interpreting alchemical symbolism and even went so far as to base his plan for an alchemical laboratory on the arcane *monas hieroglyphica*, a sort of alchemical hieroglyph composed by the famous Elizabethan magus John Dee. Needless to say, Dear, like Shapin, sees no role for alchemy in helping to shape the mechanical philosophy of Robert Boyle and his followers. Both Dear and Shapin reflect the *Gedankengut* of earlier writers like the Halls, who opposed alchemy to the important and "progressive" trends of the Scientific Revolution.[29] In reality, the primary representatives of the most recent historiography have done little but proffer a new gloss on an old and outdated story. All the emphases and prejudices of the vintage histories remain embedded in this new account, with the sole exception that now we can supposedly explain this fixed and rigid picture in terms of categories drawn from a multicolored pastiche of sociology, anthropology, literary criticism, and critical theory.[30] With their emphasis on "local knowledge," "witnessing," and "matters of fact," the new synthesists have presented a modified *explanans* for the very same *explanandum* portrayed by writers such as Butterfield and the Halls. While looking upward to erect their lofty theoretical structures they seem to have ignored the fact that the sands have shifted beneath their feet.

Brave words, the reader may say, but in what way does a consideration of alchemy really change the landscape of the Scientific Revolution? Since the writings of Paolo Rossi and Frances Yates in the 1950s and

28. Peter Dear, *Revolutionizing the Sciences: European Knowledge and Its Ambitions, 1500–1700* (Princeton: Princeton University Press, 2001), pp. 27, 53. Pace Dear, the view of Libavius as a champion of open knowledge and as an opponent of alchemy has been explicitly debunked in William R. Newman, "Alchemical Symbolism and Concealment: The Chemical House of Libavius," in *The Architecture of Science*, ed. Peter Galison and Emily Thompson (Cambridge, MA: MIT Press, 1999), pp. 59–77.

29. Shapin makes his debt to Boas Hall's treatment of the mechanical philosophy quite explicit: her 1952 "Establishment of the Mechanical Philosophy" is given top billing in his bibliographical essay, "The Mechanical Philosophy and the Physical Sciences." See Shapin, *Scientific Revolution*, p. 174.

30. For a recounting of some variations on this approach, see Jan Golinski, *Making Natural Knowledge* (Cambridge: Cambridge University Press, 1998).

later, it has been known that Francis Bacon and his followers found inspiration in the work of the alchemists. The technological dream that equated knowledge and power was already prefigured in the alchemists' desire to transmute base metals into gold. Without doubt this is true, and yet it is equally beside the point. It is one thing to imbibe from alchemy a general optimism about the human ability to alter nature as Rossi (and to a lesser degree Yates) argued, and quite another thing to use alchemy to extract from nature the experimental evidence for a corpuscular theory of matter. The claims that Rossi made for alchemy and Yates for a more diffuse Hermetic or Rosicrucian tradition did not extend to the details of the mechanical philosophy. Indeed, Yates herself viewed the mechanical philosophy as the antithetical replacement and successor to the magical worldview of which she thought alchemy to be an integral member. In her analysis, the imaginative enthusiasm of the alchemists gave way, in the end, to the cold reasoning of the mechanical universe.[31]

There are high stakes, then, to the present enterprise. If alchemy had a direct and crucial input into the mechanical philosophy of the man who first popularized that term—Robert Boyle—the polar analytical categories established by the pioneer historians of science and tacitly adopted by their recent successors will have to fall. At the same time, a radically different picture of Boyle's corpuscular theory will justly make us scrutinize the underpinnings of the new historiography. If the more celebrated writings of such theoretically informed historians as Shapin and Dear devolve into highly traditional accounts of the Scientific Revolution in their surveys, one may fairly ask whether their theory has taught us anything new, or whether it is not, in the end, mere ornament applied to a reassuring tale grown old in the telling. Innovative history does not contract itself so amicably to fit the preexisting blueprints of tired traditions. Or to use a metaphor that would have pleased the alchemists who are the subject of this book, the distillation of fresh wine does not yield vinegar.

---

31. Paolo Rossi, *Francesco Bacone: Dalla magia alla scienza* (Bari: Laterza, 1957), pp. 54–62. Frances Yates, *The Rosicrucian Enlightenment* (London: Routledge and Kegan Paul, 1972), p. 113 and passim; see also her "The Hermetic Tradition in Renaissance Science," in *Art, Science, and History in the Renaissance*, ed. Charles S. Singleton (Baltimore: Johns Hopkins University Press, 1967), pp. 255–274. For a critique of the view that alchemy was an integral member of a coherent group that one may refer to as "the occult sciences," as Yates believed, see Newman and Grafton, "Introduction: The Problematic Status of Astrology and Alchemy in Premodern Europe," in Newman and Grafton, *Secrets of Nature*, pp. 1–37, especially pp. 14–27.

A PROJECT TO INTEGRATE CHYMISTRY INTO THE HISTORIOGRAPHY
OF THE SCIENTIFIC REVOLUTION

In order to facilitate the repositioning of alchemy within the historiography of the Scientific Revolution, it may be useful to say a few words
about the development of my own research project over the last decade
and a half. In 1991, I published an edition and translation of a text
that was arguably the most influential alchemical treatise of the Middle Ages, the *Summa perfectionis* of Geber. Given the huge popularity of
this highly Aristotelian text from its inception in the thirteenth century
and throughout the Middle Ages, one might reasonably have expected
it to represent the very essence of scholastic matter theory. Traditionally
historians have depicted the medieval schoolmen as upholding a natural philosophy based on hylomorphism—the interaction of immaterial
forms that imparted qualities to an otherwise undifferentiated prime
matter. The innovators of the Scientific Revolution, on the other hand,
such as René Descartes, Pierre Gassendi, and Robert Boyle, dispensed
with hylomorphism and employed a corpuscular theory using the interactions of small, relatively unchanging bits of matter to explain the
seemingly fundamental changes that greet our senses on a regular basis.
Actions that ranged from the baking of bread to the formation of solar
systems could now be explained simply as the rearrangement of microscopic particles rather than relying on a mysterious origination and
equally perplexing departure of matterless forms.

But several things about the *Summa perfectionis* seemed to me to challenge the starkly discontinuist historical picture presented by traditional
writers on the Scientific Revolution. Unlike the dominant hylomorphic
matter theory of the Middle Ages, with its emphasis on the activity
of immaterial forms on brute matter, the *Summa perfectionis* contains a
comprehensive theory of mineral formation, chrysopoeia, and artisanal
laboratory operations expressed in terms of particles and pores. This
corpuscular orientation did not mean that the *Summa perfectionis* was
antithetical to the matter theory of Aristotle, however. To the contrary,
Geber developed a corpuscular side of Aristotelian matter theory that
is present in book 4 of the Stagirite's *Meteorology*, a work that is more
empirical and less given to abstract theorizing than much of Aristotle's
oeuvre. Perhaps because it neglected the highly metaphysical approach
to matter employed by many scholastics, the *Summa*'s corpuscular theory would still be invoked widely in the "golden age" of alchemy, the

sixteenth and seventeenth centuries, where it resurfaced in a multitude of writings on chymistry and natural philosophy.

Beyond its important corpuscular theory, the *Summa perfectionis* also contained several other striking innovations that would bear significant fruit in later centuries. Basing himself on a liberal interpretation of Aristotle's work dealing with material composition—mainly the Stagirite's *Meteorology* and *De generatione et corruptione*—Geber made the argument that humans can replicate natural products by processes that partly imitate natural ones and partly circumvent them. Although many earlier alchemists had made this claim, Geber put a new emphasis on the need to replicate natural processes in order to ensure the exact duplication of a natural product. Like his matter theory, Geber's arguments about the relationship of art (or technology) to nature would produce important results in the Scientific Revolution, where they would be recapitulated and embellished by the likes of Sennert and Boyle, among countless others. Indeed, the alchemical view that humans could perfect natural products without changing their essentially natural character provided an important element of Francis Bacon's early modern apologetics in favor of the position that artificial and natural products differ not "in form or essence, but only in the efficient."[32] A third area in which Geber broke new ground lay in his claim that of the two traditional alchemical principles, mercury and sulfur, mercury was by far the most important. This view, sometimes called "the mercury alone theory," portrayed sulfur as at best a fixing agent that could congeal the essential mercurial substance of a metal, and at worst as a corruptive, inflammable impurity. Coupled with Geber's view that the alchemist should imitate natural processes wherever possible, the mercury alone theory meant that the alchemist should focus his transmutative efforts on quicksilver, attempting to purify it and fix it within the base metals, so that they would attain the specific gravity as well as the other physical and chemical characteristics of the precious metals. During the early modern period, the mercury alone theory spawned what Lawrence Principe and I have called the "mercurialist" school of alchemy, which rejected the attempt to find the philosophers' stone in material such as niter, vitriol, and a host of other less palatable substances in favor of quicksilver.

32. Francis Bacon, *De augmentis scientiarum*, in Bacon, *The Works of Francis Bacon*, ed. James Spedding, Robert Leslie Ellis, and Douglas Denon Heath, 14 vols. (London: Longman, 1857–1874), 4: 294.

But the long-neglected field of the history of alchemy had many more surprises to offer beyond the innovations of Geber. In a subsequent book, *Gehennical Fire: The Lives of George Starkey, an American Alchemist in the Scientific Revolution* (1994; rev. ed., 2003), I showed that a largely overlooked figure in early American history, George Starkey, was actually the most significant scientific writer of the British colonies in North America before Benjamin Franklin. Following the lead of Geber and other alchemists by developing a form of corpuscular chymistry, Starkey wrote prolifically under his own name and under that of a Latin pseudonym— Eirenaeus Philalethes (a peaceful lover of truth). A devoted follower of the Flemish iatrochemist Joan Baptista Van Helmont, Starkey combined a university education at the fledgling Harvard College (A.B., 1646) with a practical formation in metallurgy at the hands of several members of the impressive ironworks facilities at Braintree and Lynn, such as William White, Robert Child, and John Winthrop, Jr. Starkey emigrated from the Boston area to London in 1650, where he immediately became a member of the informal scientific and technical circle focusing on the promotional efforts of the "intelligencer" Samuel Hartlib. Starkey soon acquired considerable fame as a prominent medical practitioner and writer on Helmontian iatrochemistry. An irrepressible advocate of chymical medicine, Starkey would go so far as to challenge the more traditional Royal College of Physicians to engage in a large-scale clinical trial, where the merits of iatrochemistry—and the corresponding failures of Galenism—would be displayed before the public. Although this medical duel never came to pass, Starkey had in the meantime attracted the patronage of the young Robert Boyle and even succeeded in giving Boyle his earliest serious training in the laboratory practice of chymistry. Under the nom de guerre of Eirenaeus Philalethes, Starkey also became the favorite alchemical author of that equally dedicated chrysopoeian, Isaac Newton. Indeed, vestiges of Philalethan alchemy may be found not only in Newton's unpublished alchemical writings but also in such well-known publications as his *Opticks* and *De natura acidorum*.

The rich legacy of George Starkey was so compelling that it produced further publications as well. In 2002, a joint effort with Lawrence Principe led to *Alchemy Tried in the Fire*, a work that built on *Gehennical Fire* and on Principe's 1998 study of Boyle and chrysopoeia, *The Aspiring Adept*. Here we were able to show that Starkey's early tutelage of Boyle was in fact Boyle's first exposure not only to chymistry but also to rigorous experimental science as such. Among the unpublished papers of John Locke, we found notebook fragments in Starkey's hand with queries interspersed in the hand of the young Boyle. On the basis of these and other unpublished documents (now available in a critical edition with commentary) we were able to demonstrate that Starkey's influence on Boyle was not an ephemeral introduction to an outdated chymistry that would soon be replaced by the more mature thoughts of Boyle as he advanced through his scientific career, but that Starkey's interpretation of Van Helmont conditioned Boyle's major discoveries in chymistry even in the latter's final years.[33] In addition, we showed that a quantitative, gravimetric tradition in alchemy, already present in the *Summa* of Geber, reached its maturity in the work of Van Helmont. The influence of Helmontian chymistry on Boyle and even on the French chymists who paved the way for Lavoisier has only begun to be acknowledged. It is clear, however, that the immensely significant discovery and articulation of "mass balance"—the equivalence of input and output weights during chemical reactions—had already been worked out by Van Helmont and elaborated into the grounds of an efficient industrial chymistry by Starkey. As we showed in *Alchemy Tried in the Fire*, this idea was transmitted from Helmontians like Starkey to the French academicians of the early eighteenth century, who in turn served as important sources of Lavoisier. In seeming recognition of this fact, Lavoisier himself paid homage to Van Helmont on several occasions, even contrasting him favorably to Boyle.

Another area in which alchemy played an important role in medieval and early modern culture lay in the shifting boundaries between the artificial and the natural, as my foregoing comments about Geber suggest. The extent to which alchemy influenced discussions of this topic up to the time of Newton and beyond has received little attention from historians. And yet even current discussions of cloning and the limits of biomedical science now being played out in the President's Council on Bioethics are still framed in terms derived from medieval and early modern alchemy. As I have recently argued in *Promethean Ambitions: Alchemy and the Quest to Perfect Nature* (2004), the discipline of alchemy was adopted by Albertus Magnus, Thomas Aquinas, and a number of other thirteenth-century theologians as a means of determining the power of art (human technology) vis-à-vis that of nature. The original theological impetus behind this discussion lay in the realm of demonology—theological writers were keen to determine the limits of demonic activity in the natural world.

33. George Starkey, *Alchemical Laboratory Notebooks and Correspondence*, ed. William R. Newman and Lawrence M. Principe (Chicago: University of Chicago Press, 2004).

Since they commonly viewed demons as being limited to the application of agents to patients by means of local motion, and this was also the realm of art, it followed that the limits of demonic power could be discerned if one knew the limits of art. As it happened, an eleventh-century attack on alchemy by the Persian philosopher Avicenna provided these medieval theologians with precisely the limit they were looking for. In the course of writing against alchemy, Avicenna had argued that art could never equal nature's products and that humans cannot really transmute the species of things. Hence metals and other products of alchemy, such as pigments and "artificial" salts, could never attain the perfection of their natural exemplars. When Avicenna's attack came to be translated into Latin, it was falsely attributed to Aristotle, "the prince of the philosophers," making it attractive to theologians bent on attaining a benchmark for demonic power. At the same time, it became de rigueur for any philosophically minded alchemist to rebut Avicenna's attack on human technology. This led to a massive scholastic disputation literature focusing on the legitimacy of artificial means for replicating natural products. The debate only intensified when in the sixteenth century followers of the iconoclastic chymist Paracelsus von Hohenheim popularized the idea that alchemists could create an artificial human being, a homunculus, within a flask. In *Promethean Ambitions*, I have argued that the ramifications of this debate extend from the Middle Ages through Goethe's *Faust* and up to modern debates about the dimensions of cloning, stem-cell research, and the possibility of ectogenesis.

There are a number of other areas currently being researched by historians where alchemy exercised important functions in the early modern period. The influence of Paracelsus, particularly in the formation of a chymically oriented medicine, has been the major thrust of Allen G. Debus's work since the 1960s, just as it was that of Walter Pagel before him. A growing cadre of European scholars, including Joachim Telle and Didier Kahn, is expanding this legacy still further.[34] The equally significant area of Paracelsian theology has received important consideration

---

34. For a representative example of Debus's work, see Allen G. Debus, *The Chemical Philosophy*, 2 vols. (New York: Science History Publications, 1977). For Pagel, see Walter Pagel, *Paracelsus: An Introduction to Philosophical Medicine in the Era of the Renaissance* (Basel: Karger, 1982). Joachim Telle's large and impressive oeuvre includes such works as his *Corpus Paracelsisticum : Dokumente frühneuzeitlicher Naturphilosophie in Deutschland*, coauthored with Wilhelm Kühlmann (Tübingen: Niemeyer, 2001) and the edited volumes *Analecta Paracelsica* (Stuttgart: Franz Steiner, 1994) and *Parerga Paracelsica* (Stuttgart: Franz Steiner, 1991). See also Didier Kahn's groundbreaking doctoral dissertation, "Paracelsisme et alchimie en France à la fin de la Renaissance (1567–1625)" (Ph.D. diss., Université de Paris IV, 1998).

from scholars such as Hartmut Rudolph, Andrew Weeks, Jole Shack-elford, and Dane Daniel.[35] Others, such as Bruce Moran, Pamela Smith, and Robert Halleux, have provided illuminating studies of the role that chymistry played in noble courts, especially German-speaking ones, in the sixteenth and seventeenth centuries.[36] Art historians such as Suzanne Butters, Thomas DaCosta Kaufmann, Michael Cole, and Lloyd De Witt have recently revealed important connections between alchemy and the visual arts in the Renaissance and the Baroque.[37] The interaction of chymistry and early modern mining technology has begun to receive its due in the work of Tara Nummedal.[38] Lawrence M. Principe has written important studies on Robert Boyle's engagement with chrysopoeia and is now engaged in further research on the role that traditional chymistry played during the Enlightenment.[39] All of these projects have thrown and continue to throw considerable new light on contributions stemming from a field that was—not too long ago—considered the epitome of folly by mainstream historians. And yet, as I have argued above, little of this material has yet been integrated into synthetic accounts of the Scientific Revolution per se, which cover the same historical period. Historians of the Scientific Revolution, particularly those who have written surveys of

35. Representative samplings include Hartmut Rudolph, "Hohenheim's Anthropology in the Light of His Writings on the Eucharist," in *Paracelsus: The Man and His Reputation, His Ideas and Their Transformation*, ed. Ole Peter Grell (Leiden: Brill, 1998), pp. 187–206; Andrew Weeks, *Paracelsus: Speculative Theory and the Crisis of the Early Reformation* (Albany: SUNY Press, 1997); Jole Shackelford, *A Philosophical Path for Paracelsian Medicine: The Ideas, Intellectual Context, and Influence of Petrus Severinus, 1540–1602* (Copenhagen: Museum Tusculanum Press, 2004); and the important recent doctoral thesis by Dane T. Daniel, "Paracelsus' *Astronomia Magna* (1537/38): Bible-Based Science and the Religious Roots of the Scientific Revolution" (Ph.D. diss., Indiana University, 2003).

36. Bruce Moran, *The Alchemical World of the German Court* (Stuttgart: Franz Steiner, 1991); Pamela H. Smith, *The Business of Alchemy: Science and Culture in the Holy Roman Empire* (Princeton: Princeton University Press, 1994); Robert Halleux and Anne-Catherine Bernès, "La cour savant d'Ernest de Bavière," *Archives internationales d'histoire des sciences* 45 (1995): 3–29.

37. Suzanne B. Butters, *The Triumph of Vulcan: Sculptors' Tools, Porphyry, and the Prince in Ducal Florence* (Florence: Olschki, 1996); Thomas DaCosta Kaufmann, "Kunst und Alchemie," in *Moritz der Gelehrte: Ein Renaissancefürst in Europa*, ed. Heiner Borggrefe et al. (Eruasberg: Minerva, 1997), pp. 370–377; Michael Cole, "Cellini's Blood," *Art Bulletin* 81 (1999): 215–235; and Lawrence M. Principe and Lloyd DeWitt, *Transmutations: Alchemy in Art: Selected Works from the Eddleman and Fisher Collections at the Chemical Heritage Foundation* (Philadelphia: Chemical Heritage Foundation, 2002).

38. Tara Nummedal, "Practical Alchemy and Commercial Exchange in the Holy Roman Empire," in *Merchants and Marvels: Commerce, Science, and Art in Early Modern Europe*, ed. Pamela H. Smith and Paula Findlen (New York: Routledge, 2002), pp. 201–222.

39. Principe, *Aspiring Adept*. Principe is now at work on a monograph on Wilhelm Homberg and his associates at the Académie royale des sciences.

the subject, seem to feel that chymistry was simply not a major contributor to the reformulation of science that heralded the birth of the modern world. My hope is that the present book, by revealing the violent rupture that alchemy helped to precipitate in traditional scholastic matter theory and by outlining the role of this discipline in the formation of the experimental version of the mechanical philosophy, will give cause for reconsideration of the traditional "grand narrative" of the Scientific Revolution. It is time to consider this topic anew rather than adding further lucubrations to the surveys and textbooks of our forebears.

The task that we have before us is not an easy one, since it requires that we reconsider both the matter theory of the High and late Middle Ages and that of the early moderns who preceded the figures with whom we are most familiar, such as Descartes and Boyle. The scholastic approach to natural philosophy made no concessions to style or context—it was dense, thorny, and replete with unstated metaphysical and religious assumptions. To the modern reader little versed in its doctrines, scholasticism is a minefield of interpretive difficulties, where one poorly understood concept can lead to a wasteland of misapprehension. It is therefore imperative that we treat this material with patience and with an eye to detail. At times this will prove difficult for the reader unused to such discussions, but there is simply no way to avoid an intimate encounter with the material that set the stage for the mechanical philosophy if we are to understand the latter as a genuine historical phenomenon rather than satisfying ourselves with the *ignis fatuus* of a rational reconstruction. In short, the reward will justify our labors, for if we are successful, the historical enterprise of attempting to transmute metals will lose its specious glimmer as a foil to rationalism and acquire a quite different sheen.

The following book therefore consists of three distinct parts. First, I consider the tradition of corpuscular alchemy in the West from its thirteenth-century inception up until the virulent debate some three centuries later in which the anti-Paracelsian Thomas Erastus and his followers attempted to debunk the matter theory of the alchemists and in turn received a stinging rebuke from the arch-polemicist and defender of chymistry, Andreas Libavius. As we will see, the scholastic alchemists were themselves Aristotelians, and yet they appealed to doctrines within the large and difficult corpus of the Stagirite that were strikingly different from those to which the equally Aristotelian Erastus made reference. The scholastic alchemists in the tradition of Geber extracted a corpuscularian Aristotle from the Greek philosopher's works, while their opponents, especially those in the tradition of matter theory descending from Thomas Aquinas, were radically opposed to this perspective. In part 2, I consider Daniel Sennert, seen against the background of this earlier alchemy as well as that of the scholasticism professed by his early modern peers. The purpose of this exposé is the elucidation of Sennert's matter theory and its goals in the light of its sources and its own gradual development over time, with the end of comparing Sennert to Boyle. I do not delve into Sennert's important contributions to medicine or the theological disputes that consumed him in his final years, both of which are topics for another study. Nor do I attempt a complete exposition of his fully mature atomism, found mainly in the 1636 *Hypomnemata physica* (Physical Dissertations), since I am more concerned with Sennert's striking transformation from a fairly orthodox representative of late Renaissance scholasticism to an outspoken if idiosyncratic advocate of Democritus. The third and final part of the book concerns Boyle's use of Sennert's theory and the experimental basis for it. As we will see, Boyle's appreciation of Sennert modified over time, but his debt to the German academic is evident even in Boyle's mature work. Boyle's explicit rejection of Sennert's hylomorphism in *Forms and Qualities* is of the highest interest in determining what the British natural philosopher viewed as his mission in life—the advancement of his own mechanical philosophy. There can be no sharper divide than the knifelike vertex separating Boyle's views on form and matter from Sennert's. And yet, their respective positions confront one another like mirror images, performing a silent pantomime where the reversal of features is in one sense real, and in another, merely apparent.

ONE

The *Mise en Scène* before Sennert

1

The Medieval Tradition of
Alchemical Corpuscular Theory

In 1619, a mild-mannered professor of medicine at the University of Wittenberg published a work intended to reconcile the warring opinions of chymists, Galenists, and Aristotelians. Despite its eirenic facade, however, Daniel Sennert's *De chymicorum cum Aristotelicis et Galenicis consensu ac dissensu* actually provided a dramatically convincing experimental demonstration that matter at the microlevel is corpuscular, thereby paving the way for the flagrantly anti-Aristotelian "mechanical philosophy" of Robert Boyle and his compatriots. All the same, regardless of its impact on Boyle and others, Sennert's demonstration was deceptively simple and relied on well-known phenomena for its power of conviction. Sennert himself derived the experiment from earlier written sources and provided little that was new in terms of empirical data. In essence, his experiment consisted of dissolving precious metals in strong acids, and then precipitating them out, seemingly unchanged, by means of alkalies.[1] This unspectacular process of dissolution and reduction was embedded, moreover, in a diffuse discussion of the history of atomism and scholastic

1. In reality, Sennert's most influential experiment involved the precipitation of silver carbonate out of a silver-nitric acid solution by means of added potassium carbonate. But upon heating, silver carbonate decomposes into pure silver, carbon dioxide, and oxygen. When Sennert heated the precipitate, he apparently (and understandably) thought that he was merely fusing particulate silver. At any rate, his major point, that the silver atoms are not decomposed by the nitric acid, still stands.

theories of matter. And yet, despite their seemingly mundane character, Sennert's operations provided a powerful basis for the increasingly experimental corpuscular theory of the seventeenth century. His experiment was sufficiently impressive that the young Boyle borrowed it almost verbatim in his first written treatment of atomism and used it in modified form throughout his many later attempts to justify the mechanical philosophy. Sennert himself employed the demonstration in many of his subsequent works to support his corpuscular theory, in contexts as varied as the discussion of occult qualities and the spontaneous generation of living beings. For him and for Boyle, the reduction of dissolved metals into their original or "pristine" state (*reductio in pristinum statum*) became a sort of crucial experiment, though Sennert himself seems to have been ignorant of the famous use that Francis Bacon made of that term.

Despite the rapid impact of Sennert's claims, his corpuscular theory stemmed from an alchemical tradition extending back in an unbroken lineage to at least the thirteenth century. The importance of this tradition, as well as Sennert's contribution, has until quite recently been largely overlooked by historians. And yet, the alchemical corpuscular theory inaugurated in the Middle Ages was of tremendous significance, for it combined the insight that matter was particulate at the microlevel with evidence for the same position acquired by means of experiment. The atomism of classical antiquity, for all its brilliance, had not originated out of an experimental context. The well-worn story of Democritus of Abdera and his teacher Leucippus reacting in the fifth century B.C.E against the monism of their Eleatic forebears is a staple of introductory philosophy courses and need not detain us here.[2] Democritean atomism was a concession to the world of the senses, of course, but its origin lay in the Abderite's philosophical desire to undermine the world of unchanging "being" proposed by his predecessor Parmenides. Democritus's brilliant solution, as every philosophy undergraduate knows, was to admit the existence of pure being in the form of atoms, but to argue that these were separated by nonbeing in the form of void spaces. Hence motion could exist, since there were gaps into which the atoms could move, and by varying the shape and size of the atoms, Democritus could in a fashion account for the multivariate complexity of the phenomenal world.

---

2. A concise history of pre-Epicurean Greek atomism with relevant primary texts may be found in G. S. Kirk, J. E. Raven, and M. Schofield, *The Presocratic Philosophers: A Critical History with a Selection of Texts* (Cambridge: Cambridge University Press, 1983), pp. 402–433.

The metaphysical origins of Democritean atomism are clear enough, even if the details of his system are lost in the haze of historical amnesia. His revivers Epicurus and Lucretius, who came at opposite ends of the Hellenistic period, made important additions to the Democritean system, but they too were strangers to the laboratory. An equally salient consideration for our story is the fact that Epicurus and his followers lived after Aristotle. Every commentator on the Stagirite's *Physics*, *Metaphysics*, and *De generatione et corruptione* felt obliged to explain Aristotle's famous rejection of Democritus. There was no similar compulsion for the scholastic natural philosophers to expand on the Epicureans. When Daniel Sennert became an outspoken adherent of atomism, he therefore announced himself to be a follower of the Abderite. And yet, paradoxically, Sennert was not a disciple of Democritus alone, but also of Aristotle. Indeed, his peripatetic tendencies reveal, better than almost any other source, the great variety hidden beneath the deceptive term "Aristotelianism" in the Renaissance.[3] Until his death in 1637, Sennert remained a serious follower of Aristotle, and yet he was also a self-styled "atomist."[4]

In order to understand Sennert's approach to atomism and the significance of his work, we must begin with the tradition of experimental and corpuscular alchemy that formed his most important source. Like Sennert himself, this tradition was highly Aristotelian in character, and yet it reflected a type of Aristotelianism that finds little or no representation in modern histories of philosophy. The peripatetic philosophy that inspired medieval alchemists was not the highly abstract, even metaphysical natural philosophy of Aristotle's *Physics* and *De caelo* but instead the more empirically oriented parts of the Stagirite's corpus, such as the *Meteorology* and certain portions of *De generatione et corruptione*. It may surprise some readers to learn that the alchemists of the High Middle Ages made a sustained attempt to conform their

3. The term "Aristotelianism" is inherently problematic, but Sennert's open and frequent appeals to the authority of the Stagirite and his insistence on hylomorphism make him as good a candidate as anyone to be labeled an Aristotelian. For some problems with the term "Aristotelianism," see Christoph Lüthy, Cees Leijenhorst, and Johannes M. M. H. Thijssen, "The Tradition of Aristotelian Natural Philosophy: Two Theses and Seventeen Answers," in *The Dynamics of Aristotelian Natural Philosophy from Antiquity to the Seventeenth Century*, ed. Cees Leijenhorst, Christoph Lüthy, and Joahnnes M. M. H. Thijssen (Leiden: Brill, 2002), pp. 1–29.

4. Sennert's atomism, as we will see, did not assume the absolute indivisibility of the atoms, either in a mathematical or in a physical sense. Despite his allegiance to Democritus, Sennert used the term "atom" to mean a small corpuscle that was difficult or impossible to decompose by operations performed in a laboratory but not necessary indestructible in an absolute sense.

art to the tenets of scholastic natural philosophy and that this approach wedded them to the debate-strewn world of medieval university discourse, but this is in fact the case. Despite modern stereotypes that cast alchemy either as the epitome of unlettered empiricism, the embodiment of dishonest greed, or the vehicle of attempts to attain a mystical union with divinity, scholastic writers on alchemy in the twelfth and thirteenth centuries were concerned above all with the attempt to fit alchemy into the rationalistic edifice of Aristotelian natural philosophy. In one sense, their efforts met with limited success, since a backlash against alchemy arose in the early fourteenth century, with mainstream theological figures, including Pope John XXII, condemning the aurific art.[5] The close connection between alchemy and important university figures of the thirteenth century, such as Albertus Magnus and Roger Bacon, found little counterpart in the fourteenth or fifteenth, and the field failed to find a home in late medieval university curricula. From another perspective, however, scholastic alchemy was a monumental success. As the present book will show, the alchemists of the High Middle Ages established an experimentally based corpuscular theory that would develop over the course of several centuries and eventually supply important components to the mechanical philosophy of the Scientific Revolution. The very movement that devoted itself singlemindedly to the destruction of Aristotelian natural philosophy was itself indebted in highly significant ways to the Aristotelianism of the Latin alchemists.

The experimental corpuscular theory of medieval and early modern Western alchemists was largely an elaboration of a textual tradition inaugurated around the end of the thirteenth century by the European author who called himself "Geber." The name "Geber" is a partial Latin transliteration of "Jābir ibn Ḥayyān," a semifabulous Arabic author who supposedly lived in the eighth century and spawned almost three thousand works. Yet the foundational text for our story—the *Summa perfectionis* (Sum of Perfection)—though dependent on Arabic models, was not a translation from Arabic but an original composition by a Latin author living around the end of the thirteenth century, probably the writer who in another work styled himself Paul of Taranto.[6] In order to

5. The text of John XXII's decretal *Spondent quas non exhibent* is reproduced and translated with discussion in Robert Halleux, *Les textes alchimiques* (Brepols: Turnhout, 1979), pp. 124–126.
6. William R. Newman, "New Light on the Identity of Geber," in *Sudhoffs Archiv für die Geschichte der Medizin und der Naturwissenschaften* 69 (1985): 76–90; and Newman, "The Genesis of the *Summa perfectionis*," in *Les archives internationales d'histoire des sciences* 35 (1985): 240–302. In

understand Sennert and later authors, we must begin with an overview of the *Summa*'s immensely influential theory and its experimental basis. The *Summa*, despite the fact that it is a highly scholastic work, presents a theory of mixture at odds with the usual understanding of Aristotle. Geber describes the combination of elementary *minimae partes* (very small particles) or *minima* that come together in a *fortissima compositio*—a "very strong composition"—to make up the two principles of metals, sulfur and mercury.[7] This theory is expressed very clearly in the twenty-fourth chapter.

> Each of these [principles] *in genere* is of very strong composition and uniform substance. This is so because the particles of earth are united through the smallest particles (*per minima*) to the aerial, watery, and fiery particles in such a way that none of them can separate from the other during their resolution. But each is resolved with the other on account of the strong union that they mutually have received through the smallest (*per minima*).[8]

Geber here asserts that the four Aristotelian elements, fire, air, water, and earth, combine "through the smallest" (*per minima*) to form the compounds of mercury and sulfur. He views the four elements as minute corpuscles that bind together to form larger complex corpuscles, united in a "very strong composition." The term "very strong composition" (*fortissima compositio*) is highly revealing, as the usual Latin word employed by the scholastics for a mixture of ingredients was *mixtio* or *mixtura*, from the Greek *mixis*. In order to understand the meaning of Aristotelian *mixis*, the contemporary reader must make a conscious effort to forget the terminology of modern chemistry, which refers to mechanical juxtapositions of particles as "mixtures" and distinguishes such uncombined ingredients from those that have entered into a "chemical compound" joined by "chemical bonds." The language employed by chemists today provides an almost exact reversal of the terminology used by Aristotle, for whom "mixture" meant a homogeneous combining of ingredients and "compound" or "composition" meant a mere juxtaposition of uncombined parts. Aristotle had claimed in his *De generatione*

---

order to avoid confusion, in the present book I will call the author of the *Summa perfectionis* "Geber," since that is how the alchemical tradition generally refers to him.

7. William R. Newman, *The "Summa perfectionis" of Pseudo-Geber: A Critical Edition, Translation and Study* (Leiden: E. J. Brill, 1991), p. 322.

8. Newman, *Pseudo-Geber*, p. 663. I have modified the translation slightly in order make the meaning clearer in the context of the present chapter.

et corruptione that genuine *mixis* occurred only when the ingredients of mixture acted upon one another to produce a state of absolute homogeneity. Otherwise, he asserted, a sufficiently keen-sighted person, such as the classical hero Lynceus, would be able to see the heterogeneous particles that made up what had seemed to be a genuinely uniform substance. Aristotle's predecessor Empedocles had of course espoused precisely the sort of theory that Aristotle was here debunking. Empedocles had maintained a century before Aristotle that the four elements were composed at the microlevel of immutable particles, which lay side by side to form compounds (what chemists today would call "mixtures"). Aristotle argued that such corpuscles could only form an apparent mixture, like wheat and barley in a jar: he dubbed such illusory mixture *synthesis*—literally "setting together." The exact Latin equivalent for Aristotle's *synthesis*, as employed in the common medieval translations and adopted by eminent scholastics such as Albertus Magnus, was *compositio*, again literally "setting together" or "putting side by side."[9]

It appears, then, that the author of the *Summa perfectionis*, who was himself trained in the philosophy of the schools, was implicitly erecting a theory of matter at odds with the concept of mixture laid out in Aristotle's *De generatione et corruptione* when he employed the expression *fortissima compositio*. As we will see in a later chapter, the *Summa's* theory was not anti-Aristotelian per se, since it derived mainly from another part of Aristotle's weighty corpus, namely, book 4 of the Stagirite's *Meteorology*. But for now let us consider the *Summa's* theory on its own terms. Geber's "very strong composition" was not a mixture at all in the strict sense of Aristotle's *De generatione et corruptione* but rather a corpuscular juxtaposition like that of Empedocles. Unlike Empedocles, however, Geber incorporated the key notion of compositional *stages* into his system: the four elements could combine to form the larger complex corpuscles of mercury and sulfur, and these in turn combined to form the corpuscles of the different metals as such. Although Geber's hierarchical stages do not map precisely onto the modern view of atoms and molecules, it is

9. Aristotle, *De generatione et corruptione*, in *Aristotelis opera cum Averrois commentariis*, vol. 5, *summa* 3, chapter 1, f. 381v. See also Aristotle, *De generatione et corruptione translatio vetus*, ed. Joanna Judycka, in *Aristotles latinus*, vol. 9.1 (Leiden: Brill, 1986), p. 68. Albertus Magnus, *De generatione et corruptione*, in *Opera Omnia*, vol. 2, book 2, tract 2, chapter 14, p. 59: "Quicunque enim dicunt elementa non generari ex adinvicem, sicut Emped. illi etiam non dicunt, quod unumquodque elementorum fiat a qualibet parte mixti corporis, sed dicunt, quod elementa ex mixto corpore, sicut lapides & lateres ex pariete qui compositi sunt in pariete, & non mixti: & hoc dictum eorum est inconveniens . . . talis enim mixtura quae non vera mixtura, sed compositio est."

not too much to view his notion of a *fortissima compositio* joining discrete corpuscles as having a kinship with the chemical bond of contemporary chemistry.[10]

The experimental basis of Geber's claim lies partly in the laboratory process of sublimation, particularly the sublimation of mercury and sulfur. Although he views these two substances as principles of the metals, they do not acquire the hypothetical quality in the *Summa* that they often do in early modern alchemy. When Geber speaks of sulfur and mercury here, he means common brimstone and quicksilver—the sulfur and mercury of the modern periodic table. Geber's claim for the corpuscular nature of these substances is based on two observational facts. First, the sublimed mercury and sulfur collect in the "aludel" or sublimatory vessel as tiny droplets (mercury) or minutely divided powder ("flowers" of sulfur)—hence the process of sublimation seems to reveal their particulate structure to the naked eye. Second, and more important, these two substances can be sublimed intact such that they leave little or no residue in the bottom of the aludel. This is the point of Geber's comment that mercury and sulfur are "resolved"—here meaning "sublimed"—without decomposing into their elementary components. As he says at another point in the *Summa*, "We see a manifest example (*experientia*) of this in the sublimation of spirits. For when a sudden resolution comes about in them by means of sublimation, the humid is not separated from the dry, nor the dry from the humid so that they be divided into the parts of their mixture."[11] Geber elaborates on this point later in the text, observing that sulfur "has a very strong composition, and is likewise uniform and homoeomerous in its particles, because it is homogeneous. Thus its oil is not borne away from it by distillation, as it is from other things having oil."[12] This resistance of sulfur and mercury to the analytical power of heat is largely due to the "very strong composition" (*fortissima compositio*) by which their elementary corpuscles are conjoined. It cannot be overstressed that Geber's claims about the substantial integrity of

10. I am not the first scholar to note the significance of Geber's *fortissima compositio*. See Kurd Lasswitz, *Geschichte der Atomistik* (Hamburg: Leopold Voss, 1890), vol. 1, 226–227; and Reijer Hooykaas, "The Experimental Origin of Chemical Atomic and Molecular Theory before Boyle," *Chymia* 2 (1949): 65–80. The crude empiricism that Hooykass imputes to alchemical corpuscular theory is clearly misguided, however. Hooykaas was unaware of the philosophical tradition of corpuscular theory stemming from Aristotle's *Meteorology* 4. See Hooykaas, "Experimental Origin," pp. 71–73. See also Henk Kubbinga, *L'histoire du concept de "molécule"* (Berlin: Springer, 2002).

11. Newman, *Pseudo-Geber*, p. 646 (Latin, pp. 280–281).

12. Newman, *Pseudo-Geber*, p. 666 (Latin, p. 328).

mercury and sulfur are based on the fact that these materials resist analysis when subjected to laboratory operations such as sublimation.[13] Geber's reliance on a laboratory process to determine the practical limits of analysis and, hence, to establish the constituents of other bodies by means of experiment would have profound resonances in the later history of chemistry. We will consider this important point more deeply later.

While focusing on the "very strong composition" of the principles, the *Summa* also refers to another factor responsible for their durability—their *uniformis substantia* (uniform substance). What precisely does he mean by this uniform substance? Aristotle had famously argued in *De generatione et corruptione* (328b 22) that the ingredients of a mixture undergo a *henōsis* or unification during the process of being mixed. Hence all the spatial regions of a true Aristotelian mixture are "homoeomerous"—identical in all their parts (*De gen. et corr.* 328a 10–12). Obviously, the type of unification envisioned in *De generatione et corruptione* is excluded by the *Summa*'s compositional matter theory. Nonetheless, the *Summa* refers to sulfur as being "homoeomerous in its particles, because it is homogeneous," and the term "homoeomerous" is employed earlier in the text to describe the principles:

> A true mixture of the dry and humid so that the humid be tempered by the dry and the dry by the humid, and so that this become one substance homoeomerous in all its parts, and temperate between hard and soft, and extensible in contusion, does not occur except by continual mixture of the viscous humid and the subtle earthy through the smallest particles (*per minima*).[14]

13. Geber was aware of the fact that sulfur will burn in the open air, of course. He even refers to abortive attempts to calcine sulfur, pointing out that they result in a 97% loss of the material by weight. Apparently he did not view the conversion of sulfur into an invisible gas (SO_2) as an analytic process. Although he had no way of knowing that a compound was being created during its combustion, the fact that the burning sulfur left little appreciable residue meant to Geber that it was not being divided into its components. See Newman, *Pseudo-Geber*, p. 666, Latin, p. 328. Elsewhere in the *Summa* Geber does speak of "dividing" sulfur, by means of additional "dregs" that are placed in the bottom of the aludel and used to hold back the part of the sulfur that is presumed to be made of larger corpuscles than the part that is elevated. He views this division principally as removing impure "earthiness" from the pure substance of the sulfur, but even so, it is clear that he does not view the *fortissima compositio* of the sulfur as being absolute. Nonetheless, since the product of his dividing is a highly purified "sulfur" rather than a more basic constituent of sulfur (such as the four Aristotelian elements), this division does not violate his use of laboratory operations to arrive at operational elements (or rather "principles"). The same may be said for his treatment of mercury. See Newman, *Pseudo-Geber*, pp. 683–687, 691–693, and Latin, pp. 361–373, 385–390.

14. Newman, *Pseudo-Geber*, p. 645 (Latin, pp. 279–280).

Here Geber was probably influenced by a tradition that derived from the famous medieval medical school at Salerno. A Salernitan interpretation had already made the corpuscularian or atomistic move of introducing the expression *per minima* into the definition of mixture in *De genera-tione et corruptione*.[15] In the same spirit, Geber uses the terms *omniomera (homoeomera)*, *una substantia*, and *uniformis* in ways that do not correspond to the literal sense of the Aristotelian text. To Aristotle, a homoeomerous substance is one that has undergone true mixture, so that every part of the substance is the same as the whole. To Geber, on the other hand, a homoeomerous substance is one where the juxtaposed particles retain their own identity but are united with sufficient cohesion that they resist the analytical agents at the alchemist's disposal (Geber's famous "very strong composition" again). In his description of sulfur, Geber equates homoeomerity with homogeneity: his sulfur is indeed homogeneous in the sense that a given sample of it must contain the same proportion of fire, air, water, and earth particles locked together in each of the sulfur corpuscles, and yet this does not commit Geber to the view that every part of the sulfur corpuscle is materially identical to the whole. Geber's concept of homoeomerity or homogeneity is therefore a relativistic one, not committing him to the absolute uniformity of Aristotelian mixture.

In addition, Geber's homoeomerity refers not only to the elemental particles in a given mercury or sulfur corpuscle but also to a multitude of mercury or sulfur corpuscles themselves. In fact, the "homoeomerity" of the two principles relates primarily to the uniform size of their minute corpuscles. The second-order corpuscles comprising the two principles, while retaining the first-order elemental particles within themselves, are very small and hence easily forced upward by the fire of sublimation. As the *Summa* puts it, "when fire rises, it always raises the smaller particles (*subtiliores partes*) with it; hence it leaves behind the larger (*grossiores*)."[16] One can therefore decompose a heterogeneous mixture into its com-ponents by means of sublimation because a weak fire will be unable to elevate the larger, heavier corpuscles. The "subtle" corpuscles will be sublimed, while the "gross" ones will be left in the bottom of the vessel. Like the ancient atomist Lucretius, Geber employs *subtilis* to mean "small" when applied to corpuscles: correspondingly, *grossus* means

15. On this Salernitan tradition, see Danielle Jacquart, "Minima in Twelfth-Century Medical Texts from Salerno," in *Late Medieval and Early Modern Corpuscular Matter Theories*, ed. Christoph Lüthy, John E. Murdoch, and William R. Newman (Leiden: Brill, 2001).

16. Newman, *Pseudo-Geber*, p. 682.

"large."[17] It is because mercury is composed of uniformly "subtle" particles that they all sublime without the deposition of much, if any, residue. To Geber, then, "uniformity of substance" and "homoeomerity" refer as much to the size of the corpuscles in a mass as they do to the material constitution of the particles considered individually.

Elsewhere, the *Summa* states that small particle size is also the cause of the great specific gravity of gold (19.31)—

> the very subtle substance of quicksilver led forth to fixation, and the purity of the same, along with the very subtle, fixed, unburning matter of sulfur, is the whole essential matter of gold ... Because it had subtle, fixed particles, its particles could therefore be much compressed; and this was the cause of its great weight.[18]

Thus gold is made of uniformly small particles of mercury and sulfur whose minuteness allows them to be packed into a constricted union, avoiding the creation of large interstices. Since the particles of mercury that make up gold are very pure, they coalesce tightly on the principle that like goes to like and do not sublime.[19] It is the absence of gross interparticular gaps that causes the great heaviness of gold when compared to the other metals. The remarkable integration between the *Summa*'s theory and experimental practice is revealed if we now consider Geber's explanation of the laboratory operation called calcination, a process that receives its name from the roasting of limestone to produce powdery quicklime. Geber defines calcination as the conversion of a thing "into powder, due to the removal of the humidity consolidating its particles."[20] This is simply the process of exposing a metal or other substance to intense heat until it becomes a dry, powdery substance or *calx*, as described in the case of iron and copper below—

> on account of their great quantity of earthiness, and large measure of burning, fleeing sulfureity, [iron and copper] are easily brought into a calx by this method. This occurs because the continuity of the quicksilver is broken, due to the abundant earthiness mixed into the substance of the said quicksilver; therefore a state of porosity is created in them, through which the sulfureity,

17. Lucretius, *De rerum natura*, book 4, lines 115 and 122.
18. Newman, *Pseudo-Geber*, p. 725.
19. For the principle of "like to like" in the *Summa*, see Newman, *Pseudo-Geber*, p. 715, n. 112. The *Summa* also accounts for the fixation of subtle corpuscles by recourse to what the author calls the *mediocris substantia* of the principles. See Newman, *Pseudo-Geber*, pp. 152–154, 164–167.
20. Newman, *Pseudo-Geber*, p. 704.

passing, can escape. Through this it is also given that the particles become rarer, and are converted into cinder because of the discontinuity due to this rarity.[21]

In his consideration of iron and copper, Geber points out that the two metals are quite porous at the microlevel, for they contain heterogeneous earthy particles that disturb the packing of their mercury corpuscles. This accords well with his earlier comments about the more closely packed substance of gold, for copper and iron are indeed lighter in specie than the noble metal, having specific gravities of 8.92 and 7.86, respectively, in comparison to gold's 19.31. As he is also aware, gold cannot be calcined by mere fire, in contradistinction to iron and copper.[22] Once again, this experimental result agrees well with his theory that iron and copper are less tightly packed than their precious counterpart, for it is precisely their internal porosity that allows them to be penetrated by the fire of calcination. When this penetration occurs, the fire drives off the earthy sulfur contained in the metal, leaving a discontinuous powder deprived of interparticular "glue."

An important corollary of Geber's compositional theory of matter is that the sulfur or mercury may be separated from various substances—especially the base metals—by simple laboratory operations such as heating and exposure to flame. The sulfurous smell given off by various ores and impure metals during their refining provides him with evidence that heat forces their unfixed (volatile) sulfur to disengage from its mercury and pass off as fine particles or vapor. The fact that calcined metals often appear in the form of yellow, red, or white powder (what we would call oxides) suggests to Geber that they also contain a fixed (nonvolatile) sulfur that remains after the volatile sulfur has been forced out by calcination. Similarly, Geber argues that some metals, such as lead and tin, contain a large proportion of unfixed mercury before their calcination. The fact that tin loses its well-known "creak" or "cry" (caused by the rubbing together of microscopic crystals upon bending a sample of the metal) after being calcined is due to loss of its volatile quicksilver. Geber proves this point by washing lead with quicksilver and then melting it, whereupon the lead gains the creak that the tin lost—as he puts it, the lead is converted to tin. The *Summa perfectionis* combines these simple laboratory processes with a battery of assaying tests to arrive at the

21. Newman, *Pseudo-Geber*, pp. 707–708.
22. Newman, *Pseudo-Geber*, p. 725.

composition of the metals in terms of varying proportions of mercury and sulfur in volatile and fixed varieties. The fact that Geber then explains cupellation, cementation, and the other assaying tests available to him in terms of his corpuscular theory shows once again the great degree of complementarity between his theory of matter and the tools of analysis provided by the laboratory.[23]

Not surprisingly, Geber goes on to depict a number of other laboratory processes, such as sublimation and distillation, in terms of the aggregation and separation of these mercurial and sulfurous corpuscles. He even explains the production of the alchemists' instrument of transmutation, the philosophers' stone, in terms of increasingly tiny mercurial particles.[24] According to Geber's view, which the historian Lynn Thorndike dubbed the "mercury alone" theory, only these subtle perfective corpuscles can penetrate deeply enough into the microstructure of a base metal to convert it into a precious one.[25] Unlike the atomism of antiquity, however, Geber's corpuscularism does not dwell on the differing shapes of the constituent particles. Instead, the *Summa* devotes most of its attention to the size of the corpuscles. *Subtiles partes*, "subtle particles," are small, volatile, and capable of penetrating deeply into narrow pores. *Grossae partes*, "gross particles," are larger (though still perhaps imperceptibly small), "fixed" or nonvolatile, and far less penetrative than their subtle counterparts. Using such variations in size as his primary differentia of particles, the author of the *Summa* was able to explain the panoply of alchemical processes at his disposal. His experimentally based theory would encounter an eager audience in the Middle Ages, and it was still given serious treatment in the first half of the eighteenth century by as illustrious a chemist as Hermann Boerhaave in the highly public forum of the *Philosophical Transactions* of the Royal Society.[26]

23. For a treatment of these experimental demonstrations of the types and relative proportions of metallic principles, see William R. Newman, "Alchemy, Assaying, and Experiment," in *Instruments and Experimentation in the History of Chemistry*, ed. Frederic L. Holmes and Trevor H. Levere (Cambridge, MA: MIT Press, 2000), pp. 35–54, especially pp. 46–49. For Geber's corpuscular explanations of assaying, see Newman, *Pseudo-Geber*, pp. 769–776.

24. Newman, *Pseudo-Geber*, pp. 162–167.

25. Lynn Thorndike, *A History of Magic and Experimental Science* (New York: Columbia University Press, 1934), vol. 3, pp. 58, 89–90, 160, 179, 624, and passim. Thorndike did not realize that Geber was the originator of the "mercury alone" theory; instead he attributed it to Arnald of Villanova. See Newman, *Pseudo-Geber*, pp. 204–208.

26. See, for example, Hermann Boerhaave, "De mercurio experimenta," in *Philosophical Transactions of the Royal Society of London* 38 (1733–1734): 145–167. I owe this reference to John Powers, who has composed an important doctoral dissertation on Boerhaave.

GEBERIAN CORPUSCULARISM AND THE REDUCTION
TO THE PRISTINE STATE

The matter theory of Geber reveals its experimental origins in another significant area as well. As I pointed out above, the probable author of the *Summa perfectionis* was Paul of Taranto, an otherwise obscure Franciscan author from southern Italy.[27] In another work, the much less well-known *Theorica et practica*, Paul presents a remarkable justification for the alchemists' claim that mercury and sulfur really do act as the ingredients of the metals despite the arguments of others to the contrary. Paul's demonstration lays out, in a strikingly prescient way, the classical "reduction to the pristine state" that Sennert and Boyle would make famous in the seventeenth century. Like the laboratory operations evoked in the *Summa perfectionis* to demonstrate the corpuscular constituents of metals and minerals, the reductions to the pristine state described in the *Theorica et practica* reveal the extraordinary integration between scholastic theory and alchemical practice embedded in the Geberian tradition.

In order for us to understand Paul's arguments fully, however, it will be necessary first to consider the antialchemical opponents whom he is trying to rebut. In essence, Paul's *Theorica et practica* takes aim at the controversial yet highly influential theory pioneered by Thomas Aquinas in the 1270s that a given substance can only receive its unity and membership in a particular species from a single immaterial entity—the so-called substantial form.[28] Here it is important to forget the vague, modern use of the term "substance," which is often used as a synonym for "material" and to think instead in scholastic categories. To the scholastics, "substance" meant that which is primary, that which can exist of itself. Accident, on the other hand, is mere attribute, that which cannot exist without substance. At the same time, "substance" implied the essential reality of a thing, as opposed to mere qualifications that could be added on to it. To say that a thing was a substance, say "bird," implied a unified group of characteristic features, including wings, a beak, the ability to lay eggs, and so forth that could be used to define "bird." If a given bird happened also to be red, big, and angry, this in no way impinged on its

---

27. Newman, "The Genesis of the Summa perfectionis," pp. 240–302, and Newman, "New Light on the Identity of Geber," pp. 76–90.

28. For a comprehensive treatment of the debate surrounding the Thomistic theory of the unity of the substantial form, at least in its early phases, see Roberto Zavalloni, *Richard de Mediavilla et la controverse sur la pluralité des formes* (Louvain, 1951), pp. 213–503.

substance, its "birdness."[29] All of this doctrine can be found in Aristotle's genuine works, but the scholastics went a step further when they considered how the four elements of Aristotle become the individual things that we encounter as part of our sensory experience of the world. What is it that converts fire, air, water, and earth into a bird, which can in turn lay eggs and perpetuate its species? Here the common solution was to think in term of hylomorphism, the interaction of form and matter. A particularly fundamental type of form, the "substantial form," was responsible for acting on matter to produce the diverse species of things in the phenomenal world. Without a substantial form, a given thing could not be a single substance, but either a congeries of disparate, unintegrated components that did not combine to give a single essence or a mere accident.[30] In the latter case, it would be like the whiteness that inheres in a white wall: although the wall (the substance) can exist without its whiteness, the whiteness of the wall (an accident) cannot exist independently. Like any accident, it must have a subject (a substance) in which it can inhere. Since Thomas held that a given substance could have only one substantial form, I shall henceforth refer to his belief as the "unity of forms" theory and characterize its many acolytes as "unitists." The opponents of this theory—equally Aristotelian in their outlook, but not Thomist—had already begun a concerted effort by the 1270s to show that there could be a plurality of substantial forms within a given substance: I will refer to them as "pluralists." The plurality of forms debate would occupy the best minds of the thirteenth century and would reemerge in altered form throughout medieval and early modern scholasticism.[31]

Like all scholastic Aristotelians, Thomas viewed matter as consisting of the four elements, fire, air, water, and earth. These in turn contained four "primary qualities"—hot and dry in fire, wet and hot in air, cold and wet in water, and dry and cold in earth. The pairs of these qualities along with an undifferentiated "prime matter" (*materia prima*) constituted the fundamental stage of material analysis. But the situation was more complicated than this, for Thomas's hylomorphism insisted that

29. For a sampling of Aristotle's views on substance and accident, see Aristotle *Topics* 1, 5 102b4–26, and *Metaphysics* 7, 1 1028a10–34.

30. The association between essence and unity is implied in the scholastic saying "ens et unum convertuntur." The supervenient character of accidents is encapsulated by the expression "Quidquid advenit enti substantialiter constituto ei advenit accidentaliter." See Zavalloni, *Richard de Mediavilla*, p. 253.

31. Obviously this use of "pluralist" has nothing to do with the ancient debate between the material pluralists such as Democritus and Leucippus and the Eleatic monism of Parmenides and Zeno.

Aristotelian *mixis*, the one type of mixture that led to a genuinely homogeneous product, could only occur if a new substantial form, called the "form of the mixture" (*forma mixti*) was imposed on the four elements.[32] This process occurred in a well-defined series of steps. First the four primary qualities of the elements produced, as a result of their mutual action and passion, a single medial quality preserving something of the extremes; this medial quality then provided the disposition necessary for the induction of the new substantial form, the form of the mixture. Yet in such a case, Thomas insisted, the imposition of the new form of the mixture meant that the four antecedent elements would be destroyed—the generation of the one entailed the corruption of the other. All that remained of the fire, air, water, and earth would be the primary qualities, the hot, cold, wet, and dry that had been paired within the elements before their destruction and that were somehow responsible for the dispositive medial quality that prepared the way for the form of the mixture. Even here it is not clear that the four qualities that remained were the original ones underlying the elements or rather similar ones that had been newly generated, for in general Thomas insisted that the primary qualities were accidents of the substantial form. If the substantial form itself had been newly introduced to the ingredients, then how could its accidents be the same ones that had been present before in the preexistent elements (which had now been destroyed)? As for the elements themselves, they were now present within the mixture only *in virtute* or *virtualiter*—"virtually"—as a result of the said primary qualities.[33]

One important result of Thomas's mixture theory was that there could be no intermediate forms between the *forma mixti*—the substantial form of the mixt[34]—and the Aristotelian prime matter. Hence, in order for a mixture to come into being, there had to be a "resolution" of the previous ingredients all the way up to the first matter (*resolutio usque ad essentiam materiae primae*). Only in this fashion could the substantial form inform the prime matter directly and without intermediary. Clearly

32. For Thomas's theory of mixture, Zavalloni's account must be supplemented by the classic study of Anneliese Maier, *An der Grenze von Scholastik und Naturwissenschaft*, 2nd ed. (Rome: Edizioni di Storia e Letteratura, 1952), pp. 31–35 and passim. A much inferior study to Maier's, though still useful on certain points, is Xaver Pfeifer, *Die Controverse über das Beharren der Elemente in den Verbindungen von Aristoteles bis zur Gegenwart, Programm zum Schlusse des Studienjahrs 1878/79* (Dillingen: Adalbert Kold, 1879).

33. Maier, *An der Grenze*, pp. 33–35.

34. "Mixt" is an archaic English word meaning a substance that has undergone mixture. It has the advantage over the term "mixture" that it can only refer to the mixed substance, not to the state of mixing.

the Thomistic theory of mixture left no place, then, for such entities as mercury and sulfur, the immediate ingredients of the metals according to traditional alchemical theory. While most alchemists argued that the metals were directly composed of sulfur and mercury and that these two principles were in turn made up of the four elements, the Thomistic theory forbade any persistent ingredients between the level of the fully formed metal with its substantial form and the Aristotelian first matter (conceived of by Thomas and his followers as "pure potency").

It is against this Thomistic doctrine that we must view the matter theory of the *Theorica et practica* of Paul of Taranto. In typically scholastic fashion, Paul begins his text with an apology for alchemy, presenting both arguments *contra* and *pro*, in a discourse that will soon put him directly at odds with the Thomistic position on mixture. Among the contra arguments in the *Theorica et practica* attacking the reality of alchemical transmutation, there are several that deny the validity of the sulfur-mercury theory of the metals. One of these arguments, probably deriving from Thomas and his followers, explicitly states that there can be no proximal principles between the four elements or the prime matter and the substantial form of a metal.[35] Paul gives this argument and its rebuttal in considerable detail, but I will abridge it in the following discussion. He begins as follows: "[T]he principles of the metals and of all mixed bodies are not other fixed principles in nature after the first elements, but... the four elements themselves, from which all generables are composed."[36] This is the thesis that Paul's unnamed opponent sets out to prove with the following argument—whenever something is

35. Strictly speaking, this part of Paul's antialchemical argument could possibly derive either from the unitists or from certain of the pluralists. The position of Thomas Aquinas postulated a "virtual" existence of the four elements, and this could be what Paul has in mind even when he speaks of a resolution up to the elements as opposed to the prime matter. On the other hand, some pluralists, such as the thirteenth-century Franciscan Richard of Middleton, accepted only a quite limited number of substantial forms under the auspices of a single *forma completiva* ("completing form"). Richard believed that man had only six substantial forms—the intellective soul, the sensitive soul, and the four elements, whereas other bodies had only the four elements and a *forma completiva*. See Zavalloni, *Richard de Mediavilla*, pp. 129–130. It soon becomes clear that Paul's major argument is with the Thomists, however, for his unnamed opponent immediately introduces the claim that during mixture there must be a reduction all the way to the prime matter and not just to the elements.

36. Paul of Taranto, *Theorica et practica*, ed. and trans. by William R. Newman, in "The *Summa perfectionis* and Late Medieval Alchemy" (Ph.D. diss., Harvard University, 1986), vol. 3, p. 31: "Praedictis autem adversari possunt duo que videntur probare principia metallorum et omnium mixtorum corporum esse quidem non aliqua fixa principia in natura post elementa prima, sed esse ipsa quatuor elementa ex quibus cuncta generabilia componuntur."

generated, it comes into being and therefore did not exist *in actu* before its generation. Hence the generation of the metals must be made out of something that no longer has actual being once the metals are generated. And so if something is generated from something else that is *in actu*, the species of the preexisting thing must be corrupted—this is what Aristotle meant by the saying *generatio unius est corruptio alterius* ("the generation of one is the corruption of another"). The argument against the mercury-sulfur theory continues thus:

> Therefore, since substantial corruption is the resolution to the first simple principles of the composite, and since the simples of all composites are the four pure elements, it seems that when a certain metal is generated, it is first resolved up to its simple elements, from which the metals themselves, just as all things, are generated. Furthermore, since there can never be said to be but one substantial form in one thing—[for] if a certain body is called such-and-such a metal as a result of a particular form, a substantial form has been generated—every preceding substantial form is necessarily corrupted, as is seen. Therefore, when a metal is generated, something else is corrupted in its substantial form by which it was in act, lest two substantial forms be said to be in the same thing at once. And since the elements themselves are in act, it seems for the same reason that a resolution of the preceding corruption should continue not only up to the elements, which would [thus] be said to remain incorrupt, but even up to the prime matter, considered not as if in act per se, but as in potentia, with all the preceding forms in it already corrupted together, and all the other succeeding forms which are said to be generated in it following continually and without intermediary.[37]

Here Paul makes a step toward linking the Aristotelian maxim that one thing's generation means the corruption of another with the Thomistic teaching of the unity of the substantial form. He continues down this

---

37. Paul of Taranto, *Theorica et practica*, in Newman, *Summa perfectionis*, vol. 3, pp. 31–32: "Quoniam igitur substantialis corruptio est resolutio ad compositi simplicia prima principia, et omnium compositorum simplicia sint pura quatuor elementa, videtur quod cum generatur metallum aliquid, prius resolvitur usque ad simplicia elementa a quibus ipsa metalla, sicut et omnia, incipiunt generari. Adhuc cum in uno nunquam dici possit esse forma substantialis nisi una—si forma aliqua corpus aliquod dicitur tale vel tale metallum, est forma substantialis generata—necesse est omnem formam substantialem precedentem esse corruptam sicut videtur. Et ideo cum generatur metallum, et quodcunque aliud corrumpitur in sua substantiali forma qua actu erat, ne simul in eodem due substantiales forme fore dicantur. Et cum ipsa elementa sint actu, etiam videtur per rationem eandem quod preexistentis corruptionis resolutio non solum usque ad elementa fieri debeat, que incorrupta manere dicantur, sed etiam usque ad primam materiam, consideratam quidem non ut per se actu, sed ut potentia, iam simul corruptis in ea formis omnibus precedentibus, et in ea continue sine medio succedentibus sequentibus formis aliis que generari dicuntur."

line, saying that since there can never be more than one substantial form in a substance, and since a metal is a substance, every previous substantial form must have been corrupted when the metal was generated. Hence when a metal is generated, there must have been a preceding resolution up to the four elements, or if these are themselves substances existing in act, then up to the prime matter itself. Only in this way can the substantial form of the new metal inform the prime matter directly and "without intermediary," that is, without other forms intervening between it and the prime matter. Hence, Paul concludes his opponent's attack on alchemy:

> It seems necessary that no preceding fixed nature should be assigned as a principle of the metals, except either the four elements or the prime matter itself, and this is the position of certain men very famous today, concerning the generation of whatever.[38]

Clearly if "no fixed nature" other than the elements or the prime matter can be considered a principle of the metals, then mercury and sulfur are excluded as candidates. Hence Paul's antialchemical opponent has invalidated the sulfur-mercury theory by invoking the Thomistic denial of a plurality of substantial forms in a given substance. The corruption of a preexistent substance that makes it possible for a metal to be generated implies that no substances—and hence no principles—can exist between the primordial ones supplied by the four elements or first matter and the metal itself.

I will ignore most of Paul of Taranto's rebuttal to his unnamed unitist adversaries' argument against the sulfur-mercury theory except to point out that he responds by maintaining the coexistence of several substantial forms within the same subject. Here Paul is clearly linking himself to those opponents of Thomas who upheld a plurality of forms. The most interesting of Paul's arguments rely on examples. Glossing Aristotle's *De anima* (414b29–32), Paul says that when the unit is subtracted from a quaternary, the ternary emerges. The smaller number always existed within the larger one "in a certain fashion" (*quodammodo*), in its own species. Similarly, when a tree is cut apart it loses its vegetative soul, but the substantial form of the wood still remains and must have been there

---

38. Paul of Taranto, *Theorica et practica*, in Newman, *Summa perfectionis*, vol. 3, p. 34: "[V]idetur necessarium quod nulla precedens natura fixa assignari debeat pro principia metallorum, nisi vel ipsa quatuor elementa vel ipsa prima materia: et hec est positio quorundam nominatissimorum hodie de generatione quorumlibet."

all the time.[39] In the same fashion, mercury and sulfur persist beneath the substantial form of a complete metal, and the sulfur, at least, can be revealed by the metal's dissolution. This principle appears from the sulfurous stench and smoke when metals are calcined at an intensely hot fire, as also by the color of their calces: clearly Paul supports the legitimacy of analysis by means of fire.[40]

The interesting integration of theory and practice that we witness here between the plurality of forms theory and the experience of the laboratory is extended quite remarkably in Paul's subsequent comments. An extraordinary chemical insight allows him to argue that the existence of intermediate principles can be proven by means of laboratory operations:

> This is expressly proven by certain experiments of this art (*experientias huius artis*), for all metals and minerals are incinerated and calcined in their own ways, as if by the resolution of their substance they are reduced to the nature of earth. But then they are resolved by techniques of art into a water, then into air through vapor and smoke, and presently through the resolution of their smoke they are reduced to the nature of water; then they are solidified by cooking into a powder or earth, and finally, having been fused by a strong fire, they return to their own original nature of whatever mineral body or metal. But if there were a complete resolution to the simple elements and not to certain mineral or metallic principles which are nearer than the first simple bodies, the metal or such and such a body would no more return from them upon [its exposure] to fire than anything else made up of the simple elements, and gold would no more return from gold than would stone or wood [return from gold], especially since fire is a common agent, behaving alike towards all and each. But since these [metals and minerals] return just the same as before, it is manifest that they were only resolved to certain components of theirs and not to the simple elements or to the prime matter, as those foresaid [philosophers] mistakenly assert.[41]

39. Paul of Taranto, *Theorica et practica*, in Newman, "The *Summa perfectionis*," vol. 3, 38.
40. Paul of Taranto, *Theorica et practica*, in Newman, "The *Summa perfectionis*," vol. 3, 30.
41. Paul of Taranto, *Theorica et practica*, in Newman, "The *Summa perfectionis*," vol. 3, pp. 40–41: "Hoc autem expresse probatur per quasdam experientias huius artis, nam metalla et mineralia omnia suis incinerantur et calcinantur modis, quasi per resolutionem eorum substantie reducantur ad terre naturam. Deinde autem per artis ingenia resolvuntur in aquam, deinde in aerem per vaporem et fumum, moxque per ipsius resolutionem fumi reducuntur in aque naturam, et deinde per decoctionem solidantur in pulverem sive terram, ultimo vere per ignem validum fusa redeunt ad propriam naturam priorem cuiuscunque mineralis corporis sive metalli. Si igitur plena esset resolutio ad elementa simplicia et non ad principia quedam mineralia seu metallica que sunt citra prima simplicia corpora, iam ex eis ad ignem non magis rediret metallum vel tale sive tale corpus, quam aliud de compositis ex elementis simplicibus, et non magis ex auro rediret

What is perhaps most remarkable about this passage is the way in which it foreshadows Daniel Sennert's much later use of the *reductio in pristinum statum* as a defense of atomism. Just as Sennert would argue that metals dissolved in strong acids could not be reduced back into metals if they had been genuinely destroyed, so Paul maintains that the metals could not be calcined, their calces dissolved, sublimed, and then finally reduced, if the metals had been corrupted all the way down to the level of the four elements. Although relying on a much earlier technology of dilute acids and bases rather than the powerful parting waters available to Sennert, Paul was able to make much the same argument. The initial analytical agent, identified by Paul as the fire of calcination, is a "common agent" that acts the same on all materials subjected to it. Even if materials react differently to it—since metal melts and wood burns—the action of heating per se is the same. If the metals were substances that underwent a profound dissolution into their elements upon heating, there would be no reason that the final reducing fire should then recombine those same elements to form the initial analysand instead of wood or stone. The fact that the metals could be regained unchanged was sufficient evidence for Paul, then, that they had only been dissolved into intermediate principles that still accounted for the specificity of metals. It is the same situation in the case of chopped lumber, where the substantial form of wood was preexistent beneath the form of the vegetative soul rather than being generated instantly when the tree is cut apart:

> It is impossible that [the substantial form of wood] was suddenly generated there by nature—with the previous form gradually corrupted—and not by actions and alterations of nature, but by sudden and voluntary [alterations] to which natural generation or corruption does not agree.[42]

According to Paul, nature cannot regenerate a form immediately from its previous corruption. As the scholastic upholders of the plurality of forms liked to argue, the fact that a corpse remains after someone dies is not due to the sudden imposition of a form of the human body (*forma cadaveris*)

---

aurum quam lapis vel lignum, presertim cum ignis commune agens sit, ad omnia et equaliter se habens ad singula. Quoniam autem redeunt hec eadem sicut prius, manifestum est ea ad quedam sua componentia tantum resoluta fuisse et non ad elementa vel ad primam materiam ut mentiuntur prefati."

42. Paul of Taranto, *Theorica et practica*, in Newman, "The *Summa perfectionis*," vol. 3, p. 38: "[I]mpossibile fuerit ipsam ibi generari per naturam subito, mox priori forma corrupta et non scilicet per alterationes et actiones nature, sed per subitaneas et voluntarias, quibus non consonat generatio sive corruptio naturalis."

on the elements at the moment of the rational soul's departure. Instead, another form was there all along beneath the rational soul, and this accounted for the persistence of the body.[43] Paul has merely extended this principle to the metals and minerals—a metal is not immediately regenerated from fire, earth, air, and water after it has been calcined. Such generations would occur suddenly and at the will of the operator, since they would not be carried out according to the normal course of nature that generated the body or the metal in the first place. Instead, the metal reveals its preexistent components once it loses its substantial form, and these principles, once they are juxtaposed in the refiner's furnace, acquire the "very strong composition" of the original metal.

Paul of Taranto's defense of the sulfur-mercury theory displays the same insistence on the permanence of ingredients within a mixt that we encountered in the *Summa perfectionis*. The "subtle" and "gross" particles of Geberian alchemy are here provided with a scholastic defense against the Thomistic theory of the unity of forms. It is clear that these particles are not "atoms" in the classical sense, since neither the *Summa perfectionis* nor the *Theorica et practica* claims that they are indivisible or inalterable. Nonetheless, the corpuscles of Geberian alchemy, as ingredients capable of undergoing retrieval by means of laboratory operations, clearly enjoy the status of semipermanent particles within a mixt. What has changed is the notion of the mixt itself—its homogeneity no longer entails that every part must be materially identical to every other part in the sense of the *mixis* laid out in Aristotle's *De generatione et corruptione*. Instead, Geberian homogeneity requires that identical juxtaposed particles of a metal be composed of principles that are themselves made of smaller corpuscles lodged within the larger ones and held together by a "very strong composition."

This tacit substitution of composition for *mixis* would find many adherents in the long tradition of Latin alchemy leading from the late Middle Ages to the seventeenth century. The foundational works behind the huge alchemical *corpora* attributed to Arnald of Villanova and Ramon Lull were heavily indebted to the matter theory of the *Summa perfectionis*, as was the original text underlying the works said to be written by Bernardus Trevirensis or Trevisanus. Other alchemists would go so far as to turn Geber's corpuscular theory into an outright atomism

---

43. See Richard of Middleton's argument against the *forma cadaveris* in the edition of his *De gradu formarum* printed in Zavalloni, *Richard de Mediavilla*, p. 79, ll. 16–26. See also Zavalloni's commentary on pp. 352–353.

with indivisible particles, as in the case of the obscure Frater Efferarius of uncertain date, the sixteenth-century Venetian priest Giovanni Agostino Pantheo, or the seventeenth-century author masquerading under the so-briquet of Eirenaeus Philalethes.[44] Geber's corpuscular theory did not die out with the waning of the Middle Ages but instead would prove immensely appealing to the self-styled atomists and corpuscularians of the seventeenth century. Indeed, what had been an implicit substitution of one Aristotelian concept (the modified *synthesis* of Geber's *fortissima compositio*) for another (the *mixis* of Aristotle's *De generatione et corruptione*) in alchemy became a strident opposition to Aristotle in toto once the public debunking of the Stagirite acquired its full early modern popularity. The hierarchical structure of corpuscles composed of smaller corpuscles would provide a fundamental building block to the mechanical philosophy of Boyle and would reappear in still recognizable form within the matter theory of his chymical heir, Isaac Newton. Underlying this belief was the clear conviction, inherited from medieval alchemy, that processes such as sublimation, calcination, and dissolution in corrosives provided ocular testimony to the analysis of matter into its more fundamental corpuscular constituents.

44. William R. Newman, "Experimental Corpuscular Theory in Aristotelian Alchemy: From Geber to Sennert," pp. 291–329; especially 300–306. The Eirenaeus Philalethes in question is not George Starkey, but another writer using Starkey's pseudonym. See William R. Newman, *Gehennical Fire: The Lives of George Starkey, an American Alchemist in the Scientific Revolution* (Chicago: University of Chicago Press, 2003), p. 268, no. 19.

2

Erastus and the Critique of Chymical Analysis

One thing that appears very clearly from the work of Geber is his insistence on the fact that laboratory operations can reveal the fundamental components of matter by means of analysis. This surprisingly controversial alchemical view would go on to form a central tenet of the system advocated by the notorious Swiss lay preacher and medical writer Paracelsus von Hohenheim (1493–1541), who saw the fundamental alchemical process as "Scheidung" or separation. Paracelsus envisioned processes ranging from the digestive system's separation of nutrient from excrement to the creative act of God himself in terms of distillation and the removal of slag during the refining of metals. He even created a neologism for the discipline of alchemy—*spagyria*—which he seemed to equate with this process of *Scheidung*.[1] By no means did Paracelsus have an explicitly corpuscular matter theory in mind, and this alone makes it unnecessary here to delve into the details of his system. Nonetheless, the Paracelsian insistence on the possibility of retrieving initial constituents by means of analysis flew in the face of scholastic theories that denied their ability to remain in a mixture. In fact, this dogged belief in the persistence of ingredients must be seen as one of the major issues that

---

1. *Spagyria* is defined thus in Paracelsus's *Opus Paramirum* vol. 9, p. 55: "darumb so lern alchimiam die sonst spagyria heißt, die lernet das falsch scheiden von dem gerechten." For *Scheidung*, see Walter Pagel, *Paracelsus: An Introduction to Philosophical Medicine in the Era of the Renaissance* (Basel: Karger, 1958), pp. 135–136, 144, and passim.

distinguished the Paracelsian view of matter from that of his scholastic opponents, despite the fact that this particular sticking point has received little comment from modern scholars.[2] The Paracelsian doctrine that the ingredients undergoing separation by means of *spagyria* were semipermanent components of material substances would in turn combine with Geberian alchemy at the end of the sixteenth century to form a central pillar of chymical corpuscular theory.

Despite its immediate appeal to the modern mind, the chymists' belief that laboratory operations could separate the preexistent components of a given substance was far from unproblematic. As heirs to Lavoisier and Dalton, we are thoroughly inured to the theory of chemical atomism, according to which the elements are fully recoverable from their compounds. But to the natural philosophers of the sixteenth and seventeenth centuries, such things were neither obvious nor simple. Even a scientist of William Harvey's stature was willing to deny that "natural bodies are primarily produced or composed of those things into which they are ultimately resolved."[3] Why was there such resistance to the now commonplace idea that laboratory processes could analyze a body into its previously hidden ingredients? On the one hand, a reliance on chymical analysis raised serious concerns about the artifactual nature of the products arrived at by laboratory operations. Scholastic natural philosophers

2. Despite the fact that Hans Kangro made a major point of the nonrecoverable status of the scholastic elements as opposed to the Paracelsian principles in his article, "Erklärungswert und Schwierigkeiten der Atomhypothese und ihrer Anwendung auf chemische Probleme in der ersten Hälfte des 17. Jahrhunderts," *Technikgeschichte* 35 (1968): 14–36, the issue receives little if any comment in Allen G. Debus's *The Chemical Philosophy: Paracelsian Science and Medicine in the Sixteenth and Seventeenth Centuries* (New York: Science History Publications, 1977). Nor does it appear in what is still the most widely used English study of Paracelsus, Walter Pagel's *Paracelsus*. Both Debus and Pagel give useful synopses of Thomas Erastus's critique of Paracelsian mixture theory (Debus, vol. 1, pp. 131–134; Pagel, pp. 311–333), but neither seems to realize that the general positions on mixture expressed by Erastus were commonplace in sixteenth-century universities. It is regrettable that the contrast between the Paracelsian insistence on retrievable ingredients and scholastic theories postulating nonretrievability has not received its due, since without an understanding of this difference, the Paracelsian theory of the *tria prima*—mercury, sulfur, and salt—looks like a mere rewriting of the Aristotelian four elements. As we will see in a later chapter, many early modern scholastics bitterly opposed the idea that analysis by fire or other means could reveal the original four elements in a mixture, precisely because the elements had lost their real being by virtue of being mixed. In such a case, the apparent fire, air, water, and earth revealed by combustion could be the same as the original elements only *in specie*, not *in numero*.

3. William Harvey, *Exercitationes de generatione animalium* (London, 1651), pp. 255–256, as quoted in Robert Frank, *Harvey and the Oxford Physiologists* (Berkeley: University of California Press, 1980), p. 19.

were keenly aware of the fact that fire is a violent agent, which could impose new changes on matter as well as separating preexistent ingredients. Although they did not generally argue that "contrived experience" or experiment is invalid simply because it imposes human "art" upon nature, therefore removing a thing from its natural course, the philosophical opponents of chymistry in the sixteenth and seventeenth centuries did assert that violent agents such as fire and intense heating could not reveal the genuine structure of matter.[4]

On the other hand, we saw in the foregoing chapter that the laboratory analysis employed already by alchemists in the High Middle Ages challenged an array of scholastic presuppositions about mixture and the nature of homogeneous substances, ideas that derived from Aristotle's concept of *mixis* versus *synthesis*. How could a chymist separate the ingredients of a genuinely homogenous substance since the very acquisition of the homogeneous state was widely assumed to mean that the diverse ingredients that had gone into it had lost their discrete character? Aristotle himself had declared the primary targets of alchemy, the metals, to be homogeneous, and yet the chymists of the sixteenth and seventeenth centuries routinely claimed to separate these substances into the three principles mercury, sulfur, and salt. Did this not prove that chymistry was a delusory activity that added new components and disguised others rather than arriving at the fundamental nature of things?

On these grounds and others, the assumption that laboratory operations could really separate the ingredients of sensible matter came under sustained attack in the famous *Disputationes de nova Philippi Paracelsi medicina* (Disputations concerning the New Medicine of Philippus Paracelsus) written by the Heidelberg professor of medicine and university rector Thomas Erastus, which was published in 1572. It would be difficult to overstate the significance of Erastus's assault to the history of matter theory—a century after the publication of his writings Boyle was still using him as an antichymical interlocutor in his own unpublished dialogue on transmutation.[5] Yet a cursory examination of Erastus's *Disputationes*

4. It is important here to avoid the commonplace belief shared by many historians of science and philosophy that the Aristotelian distinction between products of art and products of nature led scholastic natural philosophers automatically to eschew experiment on the grounds that it was "artificial" and hence could only subvert the course of nature. For a critique of this historiographical "noninterventionist fallacy," see Newman, *Promethean Ambitions: Alchemy and the Quest to Perfect Nature* (Chicago: University of Chicago Press, 2004), chapter 5.

5. Lawrence M. Principe, *The Aspiring Adept: Robert Boyle and His Alchemical Quest* (Princeton: Princeton University Press, 1998), pp. 65, 75–76, 245–252.

reveals that he adapted many of his arguments against the Paracelsian *tria prima* from a work that he had already largely written before the composition of the *Disputationes*. The work in question is Erastus's *Explicatio quaestionis famosae illius, utrum ex metallis ignobilioribus aurum verum & naturale arte conflari possit* (Explanation of the Famous Question Whether True and Natural Gold Can Be Fabricated from the Baser Metals by Means of Art), a debunking of alchemy, which accompanied the first printing of the *Disputationes*. What is most interesting about the *Explicatio* is the fact that Erastus wrote much of it before he had any deep knowledge of Paracelsian chymistry.[6] Hence most if not all of the *Explicatio*'s arguments are directed at traditional transmutatory alchemy—especially that of Geber, whom Erastus variously calls the "idol," "god," and "master of masters" of the alchemists—rather than at Paracelsus.[7] As a result, the *Explicatio* makes virtually no mention of the *tria prima*, focusing instead on the much older alchemical theory that the metals and minerals are composed of sulfur and mercury in varying degrees of purity. All the same, Erastus spends considerable energy on debunking the results of analysis by fire, since this was already used by medieval alchemists as evidence for the reality of their theory of metallic composition.[8]

It would be difficult to do full justice to the density and repetition of Erastus's *Disputatio*, not to mention the scabrous tone of invective that permeates the work like the air in a sealed office suffering from "sick-building syndrome." Yet Erastus does manage to bring down the full weight of his scholastic erudition in an all-out attack on alchemy. In this assault Erastus fuses the issues of art versus nature and the Aristotelian theory of mixture in a way that makes them seem at times to have undergone the very *mixis* described by the Stagirite in *De generatione et corruptione*. I shall try to disentangle the two issues here, however, and will therefore begin with Erastus's treatment of the resolution of mixts. In a passage that inspired many later debunkings of the *tria prima*, Erastus states that metals can in fact be resolved into oils, waters, cinders, and the like by various laboratory processes. Nonetheless, he continues, no

---

6. Thomas Erastus, *Explicatio quaestionis famosae illius, utrum ex metallis ignobilioribus aurum verum & naturale arte conflari possit*, p. 63 (appendix to *Disputationes de nova Philippi Paracelsi medicina* [1572]).

7. For the idea that Geber was the "idol," "god," and "master of masters" of the alchemists, see Erastus, *Explicatio*, pp. 71, 82, and 98.

8. A good example of this medieval usage is found in the *Theorica et practica* of Paul of Taranto. See Newman, "The *Summa perfectionis* and Late Medieval Alchemy," vol. 4, pp. 23–31.

thinking chymist would agree that they are composed of these *in actu*. Otherwise would not cheese and corpses be made of worms? According to the theories of spontaneous generation to which Erastus was heir, worms could be seen as a normal decomposition product of such materials as cheese and flesh. Hence the worms were potentially present (*in potentia*) in a given sample of cheese or flesh, but they were not, of course, actually present (*in actu*) in the cheese or flesh before those substances became rotten. A chymist, therefore, who states that the normal decomposition products of metals are always present *in actu* within the undecomposed metals is making a claim tantamount to the absurd view that cheese and flesh are made of worms.

Erastus then inverts his approach by considering the issue from the standpoint of initial ingredients rather than final decomposition products. Would anyone want to argue that a chicken can be reduced to an egg, or an egg to chicken blood, just because the chicken and the egg are initially made from those ingredients?[9] Even if we admit that a certain initial product naturally leads to another final product, it does not follow that the initial one still exists within the final one as an actual ingredient. Thus Erastus establishes that it is not necessary for a thing to be reducible into the components out of which it was made, hence taking the force out of the fire analysis by which chymists hope to demonstrate that metals are composed of sulfur and mercury. Here, however, he must confront a well-known scholastic aphorism used by the alchemists, namely *in quae dissolvi possunt composita, ex iisdem coaluerunt:* "The things into which composites can be dissolved are the things out of which they are made."[10] This maxim is based on *De caelo* 3 302a15–18, where Aristotle says (in Guthrie's translation), "Let us then define the element in bodies as that into which other bodies may be analysed."[11] Since Erastus unfailingly portrays himself as a staunch defender of Aristotle, he must therefore come to terms with the unfortunate fact that Aristotle seems here to be in agreement with the chymists.

---

9. Erastus, *Explicatio*, pp. 111–112, 105.

10. Erastus, *Explicatio*, p. 35. For parallel passages in Erastus's *Disputationes de nova Philippi Paracelsi medicina*, see Pagel, *Paracelsus*, p. 320. See also Debus, *The Chemical Philosophy*, vol. 1, pp. 133—134.

11. See the parallel treatment of this Aristotelian *locus* in Daniel Sennert, *De chymicorum cum Aristotelicis et Galenicis consensu ac dissensu* (Wittenberg: Schuerer, 1619), p. 283. Sennert renders the maxim as "Ex ijs corpora naturalia constant, in quae resolvuntur" and attributes it to "Arist., 3 De coelo, cap. 3, text 35," and "Galen, lib. 1.c.3 de elementis." In Bekker, notation the Aristotelian passage cited is *De caelo* 3 302a15–18.

ERASTUS AND THE "REDITUS PRINCIPLE"

How then to drive a wedge between Aristotle and the alchemists? Erastus invokes another scholastic aphorism of equally Aristotelian pedigree to topple the alchemists from their philosophical equanimity. The principle to which he resorts is *Non datur a privatione reditus ad habitum in naturalium generationibus*—"a return from privation to a habit [or form] in natural generations is not conceded."[12] Although already found in the thirteenth-century works of Thomas Aquinas and Roger Bacon, this rule would become one of Erastus's favorite weapons in his war on chymistry and would acquire a new prominence in the subsequent debate as a result of his writings.[13] The aphorism is based partly on *Metaphysics H* 1044b34–1045a6, where Aristotle points out that vinegar cannot become wine again, nor a dead animal be restored to life without a total corruption to its matter. Hence the vinegar must become water before it can become wine again, and the corpse must be resolved into its components, presumably the four elements.[14] Erastus invokes this rule time and time again, so I will refer to it henceforth as the *reditus* ("return") principle.

A sense of the prominence that the *reditus* rule has for Erastus can be gleaned from the use that he makes of it to interpret other passages in Aristotle's work. He applies it, for example, to the comments that Aristotle makes about "circular" and "rectilinear" generation in the final chapters of book 2 of *De generatione et corruptione*. Aristotle there distinguishes between the cyclical generation and corruption of the elements and the linear, one-time-only generation of a man (338b6–19). A sample of water can immediately become air and the air can be returned directly to water innumerable times, but a given man can only be generated

---

12. Erastus, *Explicatio*, p. 118.

13. Thomas Aquinas, *Opera omnia curante Roberto Busa S.I.* (Stuttgart, 1980), 013(QDP)6.1sc1/2. The maxim is sometimes expressed in slightly different wording, such as *a privatione in habitum non potest fieri regressus secundum naturam*. A nice discussion of this principle, heavily dependent on Franciscus Toletus, is found in Daniel Stahl, *Axiomata philosophica sub titulis xx. comprehensa* (Cambridge: Roger Daniel, 1645), pp. 406–409.

14. The examples of vinegar and a corpse are drawn from Aristotle's *Metaphysics*, book 8, chapter 7 in the pre-Bekker division of the text (= book H, chapter 5, 1045a3–5). Cf. Agostino Nifo, *In Aristotelis libros metaphysices . . .* (Venice: Hieronymus Scotus, 1559; Minerva reprint, 1967), p. 479: Nifo takes the common position that a substance cannot return from its resolution *in numero* but only *in specie*. For the standard formulation of the "rule," see Rudolph Goclenius, *Conciliator philosophicus* (Kassell: Officina Mauritiana, 1609; Olms, 1977), p. 254. Stahl, *Axiomata philosophica*, p. 406, derives the rule itself (as opposed to the examples of vinegar and the corpse) from chapter 11 of Aristotle's *Categories* (= Bekker 10 13a31–37).

once.[15] Hence elemental generation and corruption can be seen as a circle without end, whereas human birth and death is a linear process with a definite starting point and terminus. Erastus takes this as further confirmation that there can be no immediate return from privation to a habit except in the case of the four elements considered as isolated substances. Arguing that among corruptible things only the four elements experience cyclical generation, Erastus explicitly links this position to the *reditus* principle—a return from privation to habit or form would be an instance of cyclical generation, which is only permitted in the elements. Nature permits the immediate transformational sequence of water-air-water, but not that of man-corpse-man or wine-vinegar-wine. Hence, immediately after pointing out that a chicken cannot be returned to an egg, or the egg to chicken blood, Erastus explains his position.

> A return from habit or form to privation is not conceded: nor does nature proceed in reverse, but always continues forward, and by proceeding in a sort of circle she completes mutations from the elements by means of infinitely varied mixtures and temperations. The posterior [comes] always from the prior; never does [nature] generate prior from posterior up until the first elements are arrived at by means of various alterations and corruptions. Hence there is not a circular or reciprocal generation in mixtures as there is in the elements; rather, one comes to be from another or after another without a turning back, up to the point at which the mixture is dissolved again into the elements . . . This is not only revealed by that principle *a privatione sine corruptione non dare reditus ad habitum seu formam pristinam* [a return from privation to the pristine habit or form is not conceded without corruption]—of whose truth I never see any expression of doubt—but it is also understood from other considerations.[16]

Erastus's point, of course, is that the metals can no more return to their immediate ingredients than vinegar can return to wine or a corpse to a

---

15. The conversion of water to air is a simple matter in the Aristotelian system. Since water is cold and wet and air is hot and wet, one need only replace the quality of cold with that of heat to effect the transmutation of water into air. The process may be reversed by replacing air's heat with water's cold.

16. Erastus, *Explicatio*, p. 106: "Non datur enim regressus ab habitu seu forma ad privationem: nec retrorsum vadit Natura, sed prorsum it semper: ac tanquam circulo suas absolvit mutationes ab elementis per infinite varias mistiones & temperamenta procedendo. Ex priore subinde posterius, ex posteriore nunquam generat prius: donec ad prima elementa per varias alterationes & corruptiones iterum perventum fuerit. Quocirca non est generatio circularis sive reciproca in mistis, ut in elementis, sed alia fit ex alia sive post aliam sine reflexione: usque ad eam, qua mistum rursus in elementa dissolvitur . . . non ex illo tantum principio physico (de cuius veritate nunquam dubitatum video.) A privatione sine corruptione non dare reditus ad habitum seu formam pristinam, perspicitur, verum ex aliis quoque intelligitur."

living being without first being corrupted into more basic ingredients—
the Aristotelian four elements—from which the wine or the living being
can be regenerated according to the normal course of nature. And since
the alchemists claim the immediate ingredients of metals to be sulfur
and mercury, it follows from the *reditus* principle that these components
cannot be revealed by analysis. After all, for a metal to be returned to
its initial sulfur and mercury would be no different from a sample of
vinegar being returned to the wine from which it was made, according
to Erastus. The *reditus* principle, in other words, forbids the possibility of resolving a metal into any components other than the elements
themselves and therefore robs the alchemists of the crowning proof of
their sulfur-mercury theory. As Erastus says elsewhere, "No mixed thing,
which is truly generated, can again be resolved into that from which it was
proximally generated. For the generations of such are not reversed, but
always continue forward until there is a return to the first elements."[17]
Hence the *reditus* principle prohibits the resolution of a mixt such as a
metal into its *proximal* components—its *materia proxima*—which would
be sulfur and mercury if the alchemists were correct; yet it does not prohibit the resolution of mixts into their *remote* components, such as the
elements.

As we shall see, Erastus makes a great deal of this distinction between
the proximal and remote when he comes to the alchemical transmutation of metals. Before we take up that topic, however, let us ask if Erastus
really wants to deny universally that natural mixtures can be resolved into
their proximal components. To the modern reader of Aristotle, this position might seem quite strange, given that the Stagirite himself says
in *De generatione et corruptione* at I 10 327b27–29 that "it is clear that
the ingredients of a mixture first come together after having been separate and can be separated again" (in the translation of E. S. Forster).
Hence Aristotle's ancient commentators, especially John Philoponos,
proposed experiments for separating wine mixed with water by means of
sea-lettuce, sponges, and the like.[18] But unlike Philoponos, Erastus was
of course reading his Aristotle through the spectacles of medieval and
early modern scholasticism. He was therefore the beneficiary of Arabic

17. Erastus, *Explicatio*, p. 39: "[R]es mista nulla, quae vere est generata, resolvi potest iterum
in idem illud, ex quo proxime generata fuit. Generationes enim talium non reflectuntur, sed
prorsum eunt semper, dum reditum sit ad prima elementa."

18. Frans A. J. de Haas, "Mixture in Philoponus. An encounter with a Third Kind of Potentiality,"
in J. M. M. H. Thijsen and H. A. G. Brakhuis, *The Commentary Tradition on Aristotle's "De
generatione et corruptione"* (Turnhout: Brepols, 1999), pp. 21–46; cf. p. 26, n. 22.

and Latin theories of the substantial form, a concept that Anneliese Maier tells us reached its classic interpretation only after the time of the Persian philosopher Avicenna (d. 1037 C.E.).[19] If we accept the broad outlines of Maier's work on medieval mixture theories, it appears that the scholastics reached near-unanimity on the need for a *forma mixti*—a new substantial form—which had to be imposed on the ingredients of a mixture in order to make it truly a mixture as opposed to a heap of discrete components.[20] But we must add that as soon as one accepted the need for this new form, there was a strong tendency to ignore Aristotle's distinction between *genesis* and *mixis* (*De gen. et corr.* I 10 327b6–10) and to treat the imposition of the new *forma mixti* as a generation, accompanied by a corruption.[21] The problem was further complicated in the thirteenth century, when Thomas Aquinas and his followers argued that any given substance could only possess one substantial form, a topic discussed in chapter 1.[22] If the unity of forms theory were correct, it would be difficult to argue that the substantial ingredients of the mixt could be regained after their mixture, since their own forms would have been corrupted and replaced by that of the *forma mixti*.[23] Even some scholastics

19. In discussing the common rejection of Avicenna's theory of mixture by later scholastics, Maier points out that the concept of the substantial form became much more rigid after the death of the Persian philosopher. See Maier, *An der Grenze von Scholastik und Naturwissenschaft*, 2nd ed. (Rome: Edizioni di Storia e Letteratura, 1952), p. 25.

20. Maier, *An der Grenze*, pp. 27–28. Admittedly, some followers of Averroes, such as Jacobus Zabarella in the sixteenth century, denied that this *forma mixti* had an actual being distinct from that of the four elements. Nonetheless, even in this case the elements were thought to be considerably weakened or "broken" (*refractae*) in the mixture, so that their combination itself provided the *forma mixti* (see pp. 68–69).

21. This point was explicitly made by the Coimbrans, for example, who add that Aristotle himself calls mixture a type of generation in book 4 of the *Meteorology* in "cap. de putredine." See *Commentarii collegii conimbricensis societatis iesu in duos libros de generatione et corruptione* (Lyon: Horatius Cardon, 1606), book 1, chapter 10, question 2, p. 367. A nuanced treatment of the issue is found in Jacobus Zabarella, *Libri tres de misti generatione et interitu*, in *Jacobi Zabarellae Patavini, de naturalibus rebus libri xxx* (Frankfurt: Lazarus Zetzner, 1607; Minerva reprint, 1966), columns 582–585, 599–608. As Zabarella points out, Thomas Erastus was a famous proponent of the view that mixture is a type of generation. Erastus got into an extended debate with Archangelus Mercenarius on this issue. Other opponents of alchemy, such as Nicholas Guibert, also adopted this perspective. See Guibert, *Alchymia ratione et experientia ita demum viriliter impugnata* (Strasbourg: Zetzner, 1603): "Mixtio absoluta & naturae, maxime propria est totorum per tota mistio, ac eorum, quae temperari possunt, commutatorum, & unio, & haec mistio a generatione non differt."

22. See Roberto Zavalloni, *Richard de Mediavilla et la controverse sur la pluralité des formes* (Louvain, 1951), pp. 213–266.

23. For this reason Franciscus Toletus, to mention one sixteenth-century example, evades the obvious sense of *De generatione et corruptione* I 10 327b27–29 by arguing that when Aristotle

of the pluralist school denied that a body could have more than five substantial forms—those of the four elements and a "completing form" (*forma completiva*), namely, the form of the mixture—with the exception of the human being, which was thought to have at least six (the elements, the sensitive soul, and the rational soul).[24] This position too ruled out the possibility of intermediate ingredients with their own robust reality existing between the four elements and the form of the mixture, at least in all bodies other than that of man.

ERASTUS AND THE SCHOLASTIC PLURALITY OF FORMS DEBATE

We are now in a position to pose the following questions to Erastus—are the ingredients of a mixture unable to undergo recapture for the reason that the generation of a *forma mixti* requires the corruption of the preexistent forms of the ingredients up to the elements or even to the prime matter? Put another way, does Erastus's denial of such reduction to the initial ingredients after mixture mean in effect that a given mixt does not have intermediate principles between its own substantial form and the four elements or the prime matter? Is Erastus the heir of earlier arguments stemming from the medieval plurality of forms debate that would restrict the number of substantial forms in a mixt either to one or at most to a handful? We will see in the following argument that Erastus is indeed aware of this dispute, and yet he studiously avoids naming his

---

says that the ingredients of a mixt can be regained he means only that the elements can be regenerated: "Dico etiam, quod separabilia sunt, eo quod ex mixto possunt elementa generari rursus, sicut ex elementis generatum est mixtum, quae videlicet erant in virtute forma mixti, & fiunt in actu proprio: erant inquam in actu alieno, & fiunt in actu proprio: erant in actu formae mixti eminenter, & fiunt in actu specifico proprio, & hoc dicitur separari, ut quae erant simul in virtute coniuncta fiant seorsum seiuncta in actu." From *Francisci Toleti Societatis Iesu, Nunc S.R.E. Cardinalis Ampliss. Commentaria, Una cum Quaestionibus, in duos libros Aristotelis, De Generatione & Corruptione. nunc denuo in lucem edita, ac diligentius emendata. Cum privilegio.* (Venice: Apud Iuntas., 1602), fol. 58r.

24. See Zavalloni, *Richard de Mediavilla*, pp. 129–130, for the supporting text from Richard of Middleton's *De gradu formarum*, and pp. 363–366 for Zavalloni's commentary. It is worth quoting Richard himself, since he is quite straightforward in limiting the number of substantial forms in nonhuman bodies to five (pp. 129–130, ll. 85–90, 00–3): "[U]nde dico quod super illas quatuor elementares formas in nullo mixto, cuius generatio totaliter est ab agente naturali, est nisi una forma tantum, sive illud sit mixtum citra vitam, sive planta, sive animal brutum. In homine autem oportet adhuc esse aliam propter hoc quod productio hominis completur per actionem agentis increati . . . Ex praedictis patet quod in quolibet mixto praeter hominem sunt quinque formae substantiales, scilicet quattuor formae elementares incompletae et forma mixti completiva, quae est gradus habens."

medieval predecessors or openly siding with either the unitist or pluralist party:

> It suffices now to have shown that generation properly said does not exist where the subject from which something is generated is not essentially transmuted. For if [its] parts are only composed and confounded, as the stones in a wall, [or] calx, [or] flour, this is not generation but composition (*compositio*), since generation is a mutation of the whole subject so that nothing sensible is left. For it cannot happen by any created power that two different forms inhere in one matter. For any act exists in its own peculiar subject: it cannot be thought [to exist] in another. Nor do the forms of things differ in degree as qualities [do]; they behave, rather, like numbers. For nothing can be added or subtracted to them [i.e., to the forms], because the whole species would be mutated. If this were not the case, all things would be one, and things would be distinguished by [their] qualities and accidents alone: nothing more absurd or false than this could be imagined. For in this case, the following would be true—an ass is a man under other qualities and a man is an ass garbed and covered in other accidents. In this way, all things would be that "one and the same" of Parmenides and of others against whom Aristotle disputes. It is therefore necessary that what the chymists affirm, that they resolve the metals into that same matter out of which [the metals] were immediately transmuted, is false. For two substances differing in species or genus cannot generate the numerically identical thing. The subject, moreover, from whose mutation gold, for example, was generated, does not remain the same thing after the generation of gold that it was before the gold was generated from it.[25]

The general sense of this is that the generation of a true mixt—such as gold—destroys any preexistent mixts that may have served as ingredients. On close inspection, however, the passage reveals a thorough

---

25. Erastus, *Explicatio*, pp. 38–39: "Nunc sufficit monuisse generationem non esse proprie dictam, ubi subjectum, ex quo genitum aliquid est, non est essentialiter transmutatum. Nam si componantur & confundantur solum partes, ut in muro lapides, calx, arena, non generatio, sed compositio est. Generatio namque est mutatio subiecti totius, ut sensibile nihil relinquatur. Non enim potest fieri ulla vi creata, ut formae duae differentes in una inhaereant materia. Est enim actus quilibet in suo proprio & peculiari subiecto: in alio ne cogitari quidem potest. Neque differunt gradibus formae rerum, sicut qualitates, sed instar numerorum habent: quibus nec addi nec detrahi aliquid potest, quia protinus tota mutata sit species: Nisi hoc esset verum, omnia unum essent: resque solis qualitatibus & accidentibus distinguerentur: quo nihil possit fingi falsius & absurdius. Tunc enim vera haec esset: Asinus est homo sub aliis qualitatibus: & Homo est Asinus aliis amictus & opertus accidentibus. Hoc pacto omnia unum idemque illud Parmenidae, atque aliorum, contra quos disputat Aristoteles, fuerint. His positis, falsum esse oportet, quod Chemici affirmant, se metalla in eandem illam materiam resolvere, ex qua transmutata proxime procreata fuere. Non enim potest res eadem numero ex duabus substantiis specie aut genere differentibus nasci. At subjectum, ex quo mutato aurum, exempli gratia, genitum fuit, non manet idem hoc post auri generationem, quod fuit antequam ex eo aurum fieret."

understanding of the traditional debate between those who upheld a plurality of forms in a substance and the proponents of the unity of the substantial form. First Erastus distinguishes between Aristotelian composition (*compositio*, or *synthesis* in Greek), in which the parts of a thing are merely juxtaposed, and true generation, which requires a transmutation of the whole essence, resulting in genuine homogeneity. He then elaborates on the meaning of homogeneity by saying that two forms cannot directly inform the same bit of matter, a thesis that few Aristotelians would have disputed. Even the pluralists argued that the multiple forms inhering in a substance occupied different levels of being and were subordinated to one another, rather than competing mutually to define the essence of a single portion of matter.[26] Erastus bolsters this thought by adding that every act can only be found in its own subject. In other words, there is a direct correspondence between a given substantial form and its peculiar matter—it is not as though any old form can be imposed willy-nilly on any arbitrary material subject. As Erastus says elsewhere, every substantial form must have its own properly disposed matter, which differs from the matter informed by any other substantial form, and this matter must consist only of the four elements. The primary qualities of the elements, hot, cold, wet, and dry, must act on one another in a way that produces a *temperamentum* (a tempered state) peculiar to each substantial form.[27] The four elements then provide the proximal matter in which that substantial form can inhere, not sulfur and mercury.

Erastus then invokes an argument that links him directly to the scholastic unitists of the High Middle Ages. He points out that substantial forms do not behave like qualities—they are not intended or remitted like redness or heat—but instead act like numbers. When two numbers are added together, he says, "the whole species is mutated." The origin of this argument lies in two Aristotelian passages—*De anima* II 414b 29–32 and *Metaphysics H* 3 1043b–1044a2. In the *De anima* passage, the Stagirite points out that in animals the vegetative soul is in the sensitive soul just as a triangle is within a quadrilateral. The Thomistic

26. The phrase "Non enim potest fieri ulla vi creata, ut formae duae differentes in una inhaereant materia" does not in itself commit Erastus to the unity of the substantial form, however, since it is not clear whether he is speaking of elementary or higher forms. As Maier points out, the majority scholastic opinion forbade that the four elemental forms could simultaneously inform the same bit of matter: the debate over the plurality of forms concerned mainly higher forms than those of the elements. See Maier, *An der Grenze*, p. 27.

27. Erastus, *Explicatio*, p. 110: "Quippe forma substantialis quaelibet non potest inesse alia, quam in sua quadam rite disposita materia." See also pp. 107, 117–118.

unitists took this to mean that the vegetative soul has a merely potential being with regard to the sensitive; it does not have a separate substantial form. The *Metaphysics* text, on the other hand, asserts that a definition is just like a number—just as a number is no longer the same number once someone has increased or decreased it, so an augmented or diminished definition is no longer the same definition. Again, the unitists took this to mean that the "quiddity" (literally the "whatness," that is, the essence) of a thing was not capable of augmentation or diminution. In the same way that a man could not be more of a man or a horse more of a horse, so any substantial form was present or absent in its totality.[28] This is the point of Erastus's denial "that the forms of things differ in degree as qualities [do]." The pluralists, to the contrary, with their notion that one form can be subordinated to another, often viewed a form as acting like the matter to another form—hence there was a relation of inferior to superior, and a corresponding scale of completeness (*perfectio*). The substantial form did not acquire the "all or nothing" status that the unitists bestowed on it.[29]

Following this, Erastus lays out some absurdities that would follow, in his view, if there were no substantial generation but only composition, a belief that he wants to impute to the chymists. In a world without essences, he says, there would be no difference between a man and an ass other than their accidental qualities. Since no substantial forms would inform matter, there would be but one material substrate of all things rather than four elements, and this monistic being would be distinguished by purely accidental differences. In a rather hyperbolic fashion, Erastus equates this underlying material with the immutable "one thing" of the Eleatic philosopher Parmenides, who had denied the reality of change.[30] In his view, the chymists' approach would relegate the phenomenal world to the status of an illusion.

Finally, Erastus returns to the issue that has been in the back of his mind all along—chymical analysis. Elaborating on a principle that he has possibly borrowed from the Arabic philosopher Averroes, Erastus

28. See Richard of Middleton's presentation of the unitist position in the edition of his *De gradu formarum* published by Zavalloni, *Richard de Mediavilla*, pp. 42–43, ll. 37–49 (for the argument from *De anima*) and pp. 45–46, ll. 85–96 (for the passage from the *Metaphysics*).
29. For the subordination of forms according to the pluralists, see Zavalloni, *Richard de Mediavilla*, pp. 310–315.
30. Parmenides, a strict monist, believed that the phenomenal change and multiplicity of the sensual world are merely an illusion. In reality, there is only "one thing," undifferentiated and unchanging. See G. S. Kirk, J. E. Raven, and M. Schofield, *The Presocratic Philosophers: A Critical History with a Selection of Texts* (Cambridge: Cambridge University Press, 1983), pp. 239–262.

says that two substances "differing in species or genus cannot generate the numerically identical thing."[31] The import of this objection can be clarified by considering the case of gold. If the chymical theory of metallic generation were correct, then mercury and sulfur naturally generated out of the four elements within an auriferous mine would be the source of a given sample of the noble metal. A subsequent laboratory analysis of that same sample of gold should, by the chymists' theory, reveal the preexistent sulfur and mercury within it. This, however, presents a problem, for on the just-introduced principle, two substances belonging to different genera or species cannot generate the very same portions of sulfur and mercury. And yet by the chymists' hypothesis, in the one case, the sulfur and mercury originated directly from the four elements, while in the other case, the two alchemical principles were the products of gold. Hence the mercury-sulfur theory, in Erastus's view, has been shown to lead to a reductio ad absurdum. The final statement, that the proximal material out of which the gold itself has been generated is not the same as that into which analysis reduces it, is merely a restatement of the foregoing arguments.

As we saw before, then, Erastus is indebted to the unitists for his general argument that the generation of a new substance requires any preexistent *forma mixti* within it to be corrupted. Unlike Thomas Aquinas and his immediate followers, however, Erastus repeatedly speaks of a resolution up to the four elements rather than to the prime matter. He does not explicitly insist on the immediate inherence of the substantial form in the Aristotelian prime matter, nor does he argue that mixture requires the destruction of the four elements. In this respect, he resembles some of the pluralists, who tried, as we have seen, to restrict the number of substantial forms in a (nonhuman) mixt to five—the forms of the four elements and the *forma mixti*. It would appear, then, that Erastus's position is a self-conscious attempt to find his own way by steering between the scholastic subtleties of the unitists and the pluralists and basing himself directly on the Stagirite wherever possible.[32]

31. Averroes, in *Aristotelis de generatione animalium, Aristotelis opera cum Averrois commentariis* (Venice: apud Junctas, 1562–1574; reprint, Frankfurt am Main: Minerva, 1962), vol. 6, 44v: "Et sicut non potest dari unum & idem factum ab arte, & natura, ut imaginati sunt Archymistae: cum causae artis, & naturae sint diversae: sic etiam causae entium naturalium non possunt esse diversae, & convenire in specie, & forma." If this is indeed the source of Erastus's dictum, then he has imported the additional criterion of numerical identity from elsewhere.

32. My view here is bolstered by a passage in Erastus's famous attack on Paracelsus, where the Swiss academic contrasts his method with that of the "modern reformers," who have become

THE CHYMISTS AS NAÏVE CORPUSCULARIANS

Let us return, then, to the point with which Erastus began his discussion of the persistence of ingredients—the distinction between composition (*compositio*) and generation. It is worth examining his thoughts on composition more deeply, for the concept provides a key to the Erastian complaint against chymistry. In essence, he views the chymists as naïve corpuscularians who have failed to distinguish between true mixture and mere juxtaposition. It is the chymists' failure to appreciate this distinction that has led them to their spurious belief in the analytical power of fire:

> Everything is produced from those things into which it is dissolved. Nothing is truer than this proposition, if it is rightly understood. For it is not necessary that just any thing be considered to be made up out of those components into which it is corrupted—or which are generated from it once corrupted. Otherwise all things would be made out of cinder, since all earthly bodies can be reduced into cinders . . . Therefore the proposition must be taken to concern the dissolution [of a substance] into its parts from which it is composed as matter existing within *in actu*, not *in potentia*. Clearly it cannot be gathered from [this proposition] that all things can be dissolved into their proximal matter from which they have been proximally made. On the contrary, this can only occur when a thing is glued together from certain things existing within *in actu*, as through a certain composition (*compositionem*), but not when the thing is generated through substantial transmutation of the matter.[33]

As we can see, Erastus implicitly contrasts composition to the Aristotelian concept of genuine mixture. As we have seen, Aristotle of course maintained that genuine *mixis* occurred only when the ingredients of mixture acted upon one another to produce a state of absolute homogeneity. *Compositio*, on the other hand, was the Latin word for Aristotle's *synthesis*, which meant a juxtaposition of ingredients without true mixture. It was

---

ensnared in the subtleties of traditional scholasticism. See Erastus, *Disputationum de nova Philippi Paracelsi medicina pars altera* (Basel: Petrus Perna, 1572), pp. 201–202.

33. Erastus, *Explicatio*, p. 105–106: "Res enim quaelibet ex illis conflata est, in quae dissolvitur. Hac Propositione nihil est verius, si recte intelligitur. Etenim non est necesse, ut res aliqua putetur ex illis esse constituta, in quae corrumpitur: aut quae ex ea corrupta generantur. Sic enim ex cinere facta essent omnia, cum in cineres redigi terrena corpora cuncta valeant.... Accipienda igitur propositio est de solutione in partes suas, ex quibus ut materia actu inexistente composita est, non potentia. Quippe non potest ex ea colligi, res quaslibet in materiam proximam, ex qua scilicet proxime facta sunt, dissolvi posse. Verum hoc esse tum solum potest, cum ex aliquibus actu inexistentibus, veluti per compositionem quandam res coagmentata est: non cum per substantialem materiae transmutationem generata est."

for this reason that Geber, in his *Summa perfectionis*, explicitly employed the term "very strong composition" for his corpuscular bonding of elements to form the principles of sulfur and mercury. When Erastus speaks of certain parts existing within in actu, which are glued together to form a "composition," then, he too has in mind the Aristotelian *synthesis*, or *mixtio ad sensum*, rather than genuine *mixis*. The point that Erastus wishes to make is that resolution into proximal constitutive components can only occur when no genuine mixture has taken place, but only an apparent mixture, such as that of barley and wheat shaken together in a container. In the case of genuine mixture, the *reditus* principle and the destruction of previous "forms of the mixture" prohibit any resolution other than the absolute return into the four elements. In the case where the analysand is a composition rather than a mixture, however, such a return to the original components of the apparent mixture is indeed possible.

Although we have shown that Erastus occasionally employs traditional unitist arguments, it is really the *reditus* principle that supplies him with his principal weapon against chymical analysis. Erastus's extensive use of the *reditus* principle does not end with his debunking of fire analysis. His ultimate goal, of course, is to invalidate the claims of alchemists that they can transmute one metal into another. Geber, the inventor of the widely influential "mercury alone" theory, had argued that the philosophers' stone is an ultrapure and subtle mercury that can combine with the inherent mercury found naturally in metals. This mercurial theory of the philosophers' stone obviously drew its force from the belief that the metals contained their own intrinsic mercury, to which the philosophers' stone could bond. Since the alchemists built their sulfur-mercury theory on empirical evidence of analytical decomposition such as that which we witnessed in Geber's examples, Erastus hopes to cut the legs out from under the alchemists' transmutational goals by invalidating their laboratory analysis. Additionally, a demonstration that man cannot make genuine mixts by means of his own operations would exclude the idea that alchemy can make metals (at least as long as one accepts that the metals are products of true *mixis*, as Erastus does). To this end, Erastus further develops his argument that there can be no immediate return from privation to a habit. As he says time and time again, it is impossible for two species within the same proximal genus to be mutually transmuted. The immediate object of this principle is the metals, of course, for alchemists had long argued that lead and silver, for example, represent different species within the genus of the metals. Erastus represents this view with the aid of various life forms, however. Hence he says that a worm can be transmuted into a fly or an egg into a chicken because the worm is not passing into a more noble species of worm or the egg into a better egg. The fact that less noble and more noble worms or eggs belong to the same proximal genera, namely, worms or eggs in general, respectively, prohibits any transmutation from taking place within the category of worms or eggs. Rather than becoming a more perfect worm, the worm enters the new genus of flies, and the egg too becomes the member of an altogether different genus, namely, that of fowl.

Despite the fact that Erastus adduces no direct Aristotelian support for his edict regarding transmutation, it is not difficult to see its close relationship to the *reditus* principle. We must bear in mind that for Erastus the *reditus* principle not only prohibits the continued existence of prior forms once a transmutation has occurred so that a chicken, for example, cannot become an egg; it also eliminates the possibility of a product made by corruption returning to the state immediately prior to its corruption (as in the impossible case of vinegar becoming wine). As he says elsewhere, it would be impossible for a mercury made out of corrupted gold to return to gold once it lost its initial form of aureity for precisely that reason.[34] It would be returning to the form that it had just lost, and since it has lost that form, the mercury no longer partakes of the "common and proximal matter" (*materia communis et proxima*) from which gold is made.[35] Furthermore, if this putative reduction of auric mercury into gold is impossible, it is a fortiori impossible for the mercury to become another metal, since it will lack that metal's *materia proxima* even more surely than it lacks the *materia proxima* of gold.

We may put Erastus's words in a slightly different form, as follows: in order for genuine transmutation of a mixt to occur, the substantial form of the mixt must be corrupted and replaced by another substantial form. Since there is no immediate return from privation to a habit, the new *forma mixti* cannot be the same as the old—or even belong to the same proximal genus. As Aristotle pointed out in *Metaphysics H*, vinegar cannot again become wine without passing through the whole

34. Erastus, *Explicatio*, pp. 118–119: "Aurum ponamus mutatum in hydrargyrum, . . . & quaeramus, utrum haec substantia reduci ad aurum queat, an non? Si poterit priori conditioni restitui, auri formam retinuit, accidentia duntaxat communia, quae adesse & abesse absque rei interitu possunt, amisit. Non enim restitui potest, quod forma spoliatum est: propterea quod non datur regressio, ut saepe iam est dictum, a privatione ad habitum, sive a corruptione ad formam."

35. Erastus is operating on the principle that a given substantial form cannot inhere in just any old matter, but that the matter must be disposed in a particular fashion for each form: the matter must acquire a certain *temperamentum*, presumably acquired by the interaction of the four elementary qualities. See Erastus, *Explicatio*, pp. 107–108, 117–118.

process of generation and corruption *ab initio*. To Erastus, this means that the vinegar would have to be corrupted into its elements, these would have to produce new grape vines, the harvested grapes would have to undergo fermentation, and only then would the wine be regenerated. Since the vinegar cannot immediately return to its former state of being wine, different species of wine, such as Pouilly-Fuissé and Saint-Emilion, would be equally unattainable by the vinegar. Likewise a metal, once its metallic form has been corrupted, cannot once again become a metal, be it lead or silver, unless it returns first to the primitive state of the four elements and is regenerated underground according to the ordinary course of nature.

To summarize Erastus's position, then, alchemy can neither resolve a genuine mixt into its components nor transmute one mixt into another that belongs to the same proximal genus. When alchemists appear to be doing either of these things, they are not working with genuine mixts but instead with compositions, which are mere aggregates of heterogeneous corpuscles. In such cases, it is easy to reduce the composition into its parts, since they were already present within it, *inexistentiae in actu* (existing within in act). And in the case of apparently successful transmutations, the alchemist merely adds coloring agents or other components that alter the accidental appearance of a metal without combining to form a homogeneous mixture or leading to a genuine change of species. To perform that feat, man would have to be able to impose a substantial form on matter, which would exceed the powers of art. As Erastus says,

> For these [things mixed by alchemy] are conjoined by art, which is the ape of nature: it cannot make substances: but it effects something *per accidens* from many [ingredients] artificially conjoined: such as theriac and other drugs of this sort. Indeed, if art could make one thing per se out of many, it [i.e., art] would not be an external principle but rather an internal one, concealed within and extended through the whole matter.[36]

In this fashion, Erastus rules out the possibility of man making genuine mixts by means of art. He is probably basing this view on a similar

---

36. Erastus, *Explicatio*, p. 121: "Coniunguntur enim haec per artem, quae naturae simia est: substantias fabricari nullas potest: sed ex multis inter se artificiose conjunctis unum quiddam efficit per accidens: quale est theriaca, & alia huiuscemodi pharmaca. Sane si posset ars ex pluribus iunctis, unum per se efficere, non esset ea principium externum, sed internum: intraque materiam totam abditum & extensum."

prohibition against human attempts to replicate homogeneous mixture made by Galen.[37] At the same time, however, Erastus is clearly appealing to the distinction that Aristotle erects between natural and artificial things in the *Physics* (II.1 192b9–19). There Aristotle argues that only natural things have an internal principle of change (*archē kinēseōs*), whereas artificial ones require an external agent to change them. Erastus points out that genuine mixtures manufactured by man would have an artificial essence that would permeate their entire being. In such a case, there would be no difference between products made by man and those made by nature, with the result that man would become a sort of second God. In Erastus's view, to claim that "art can give a natural form to matter" is the same thing as to say that "art creates, or is the Creator."[38] After all, the infusion of forms into matter is effectively an act of creation, insofar as it creates substances.[39] Only nature can infuse forms into matter, and since nature is simply the *ordinaria Dei potestas*—the ordinary power of God—it is impious to attempt the transmutation of substances. At times this consideration leads Erastus into a crescendo of rage, as in the following passage:

> Thus since the origin of substantial forms is from God, and the insertion of such a form into matter should be called nothing else but a certain creation, it is clear that they who assume this for themselves—namely, putting forms naturally in matter prepared in any fashion—impiously arrogate the works of divinity to themselves.[40]

The alchemists are therefore nothing but irreligious imposters who assume the power of God and wage war on nature.[41] They are "gigantic

37. Galen, *Mixture*, in P. N. Singer, trans., *Galen: Selected Works* (Oxford: Oxford University Press, 1997), p. 227.

38. Erastus, *Explicatio*, p. 75: "Primum est, artem dare posse materiae formam naturalem. Quod perinde est, ac si dicas, artem creare sive creatorem esse."

39. Erastus, *Explicatio*, p. 79: "Si aliter metalla auri forma exornare tentant, eiusmodi tentant, quae solus Deus aut natura (quae nihil est aliud, quam ordinaria Dei potestas) praestare possunt. Equidem transmutatio talis seu infusio formarum naturalium, re non dissert a quadam creatione. Quoniam non omnis creatio totum creatae rei cuiuslibet ex nihilo facit."

40. Erastus, *Explicatio*, p. 79: "Proinde cum formarum substantialium ortus a Deo sit, nec aliud dici talis formae in materiam insertio debeat, quam quaedam creatio, patet illos sibi divinitatis opera impie arrogare, quicunque hoc sibi sumunt scilicet formas naturaliter in materiam quovis modo praeparatam immittere."

41. Erastus, *Explicatio*, p. 68.

gold-destroyers (*chrysophthoroi*)" who see themselves as the equals of God and Nature.[42]

The extraordinarily vituperative response of Erastus to alchemy and chymical analysis was far from unassailable on purely scholastic grounds. His extreme position on the dichotomy between natural and artificial products was by no means the exclusive view, although it could be justified by reference to Galen and a number of Thomistic authors.[43] As for his claim that there was no possibility of retrievable intermediate substances existing between the four elements and a perceptible homogeneous mixture, this too was open to qualification. We have seen that a long scholastic tradition had opened up the possibility of a plurality of forms existing in one subject, especially for man. Some of the pluralists even went so far as to argue that there were separate forms for flesh, bone, nerves, blood, and organs.[44] Nonetheless, Erastus's application of the *reditus* rule to alchemy was surely the most exhaustive treatment of the subject that the early modern period produced, and it had long-lived consequences. The *De febribus* (1642) of the Flemish chymist J. B. Van Helmont would still invoke the *reditus* rule in this context as an argument against the possibility that pearls, coral, or silver lost their substantial form when dissolved in nitric acid—"pearls or corals persist in their own nature no otherwise than silver dissolved in *chrysulca* [nitric acid] persists safe, obviously unchanged in all its original qualities. For otherwise the same silver could not be again recovered, since there may not be a regression from privation to habit."[45]

In short, Erastus pointed to a striking dissonance between the alchemical view that metals contain their principles within themselves as fully formed ingredients and the prevailing scholastic view that these ingredients must lose their complete, individual being so that they can merge to form a fully homogeneous substance. The irascible opponent of chymistry clearly saw that the "idol" and "god" of the alchemists, Geber, was really not a proponent of Aristotelian *mixtio* but of *compositio*. The strategy of Erastus, therefore, was to argue that the alchemists

---

42. Erastus, *Explicatio*, pp. 67–68.

43. See Newman, *Promethean Ambitions*, chapter 2.

44. Zavalloni, *Richard de Mediavilla*, p. 280 and n.60 on the same page.

45. Joan Baptista Van Helmont, *Febrium doctrina inaudita*, in *Opuscula medica inaudita* (Amsterdam: Elsevier, 1648), p. 38: "Nam perlae aut coralla non secus in pristina sui natura adhuc perstant, quam alioquin argentum in Chrysulca dissolutum sospes perstat, cunctis suis pristinis qualitatibus plane immutatum. Nam alias idem argentum, inde iterum non posset reperi, cum a privatione ad habitum non detur regressus."

did not adhere to a theory of mixture in conformity with Aristotelian *mixis*: their claims were intelligible only on the assumption that metals, minerals, and other materials were composed of heterogeneous ingredients. And yet the Geberian theory was largely based on a tradition that found its origins in book 4 of Aristotle's *Meteorology*, a work in which the Stagirite proposed a variety of corpuscular explanations for material change.[46] The energetic attacks that Erastus made on analysis performed by laboratory methods therefore produced a deeply ironic result. Since Erastus himself agreed that chymical analysis made sense in terms of a corpuscular theory of composition, he provided an indirect incentive to the adoption of that position. One needed only to give up or play down the reality of homogeneous *mixis* in order to make sense of *spagyria*, of the production of metallic "mercuries," and even of the artificial making of gold by chrysopoeia. But how could one abandon *mixis* while remaining within the pale of university scholasticism? Here Erastus's attacks would have a further unforeseen consequence. They would encourage the chymically inclined among natural philosophers and physicians to find an Aristotle who approved of corpuscular theory. In short, the Erastian approach would help to foster a rediscovery of book 4 of the *Meteorology* and the corpuscular Aristotle.[47]

46. For Geber's dependency on Aristotle's *Meteorology* IV, see William R. Newman, "The *Summa perfectionis* and Late Medieval Alchemy" (Ph.D. diss., Harvard University, 1986), vol. 1, pp. 292–295, 308, 318–320.

47. In addition to the examples of Andreas Libavius and Daniel Sennert, one could adduce Gaston DuClo as an influential author who reacted to Erastian arguments by expanding on the more corpuscularian side of the Aristotelian corpus. For an overview of DuClo's corpuscular theory, see Lawrence Principe, "Diversity in Alchemy: The Case of Gaston 'Claveus' DuClo, a Scholastic Mercurialist Chrysopoeian," in *Reading the Book of Nature: The Other Side of the Scientific Revolution*, ed. Allen G. Debus and Michael T. Walton (Kirksville: Sixteenth Century Journal, 1998), pp. 181–198. One sees the same tendency even among nonchrysopoetic writers influenced by Erastus, such as Nicholas Guibert, who will be discussed in a subsequent chapter.

3

Aristotelian Corpuscular Theory
and Andreas Libavius

The famous opponent of chymistry Thomas Erastus presented his antialchemical invective in stridently Aristotelian terms, as though devotees
of the aurific art, and particularly the "idol," "god," and "master of masters" of the alchemists, Geber, were automatically to be excluded from
the pale of peripateticism.[1] There is a deceptive and unintended irony in
this, however, for in reality Geber was himself a follower of Aristotle. The
deepest roots of Geber's corpuscular theory lie in a section of Aristotle's
system that has received relatively little study from modern historians of
philosophy. I refer to the Stagirite's *Meteorology*, especially book IV. It is
well known that *Meteorology* IV lays out a detailed corpuscular description of matter expressed in terms of *poroi* (pores) and the *onkoi* (corpuscles)
that can fill them. This theory underlies the experimental corpuscularism
of the *Summa perfectionis* both as an immediate source and as the basis of
a long commentary tradition upon which Geber is dependent. In the sixteenth and seventeenth centuries, however, the *Meteorology* came under
increasing scrutiny on its own terms, and a natural concomitant of this
heightened interest was a reintegration of the Geberian tradition with

---

1. Thomas Erastus, *Explicatio quaestionis famosae illius, utrum ex metallis ignobilioribus aurum verum
& naturale arte conflari possit* (appendix to *Disputationes de nova Philippi Paracelsi medicina* [1572]),
pp. 71, 82, and 98.

its own Aristotelian roots.[2] At the same time, alchemy itself underwent a metamorphosis at the hands of Paracelsus von Hohenheim and his followers, with the result that an increased emphasis was placed on the twin processes of analysis and synthesis (*spagyria*).[3] Hence the corpuscular theory of the *Meteorology* merged with the tradition of Geberian alchemy—now seen in the light of Paracelsian *spagyria*—to yield a widely held physical theory based on the experimental analysis and synthesis of substances believed to consist of minute particles, a position that would reach its consummate expression in the work of Daniel Sennert. Despite the fact that we are unaccustomed to seeing Aristotle as a corpuscular theorist, the *syndiacritical* interpretation of his work, as it came to be called, was viewed by influential academic authors as being a natural outgrowth of his philosophy.[4]

In order to see how this could be so, we will need to begin with *Meteorology* IV. The fourth book of Aristotle's *Meteorology* has long puzzled Aristotelian scholars, such *éminences grises* as David Ross and Werner Jaeger having dismissed it as spurious. It is right to our point that the most sustained critique of the authenticity of *Meteorology* IV was made by Ingeborg Hammer-Jensen in 1915, precisely because the text seemed to her to be uncharacteristically "mechanistic."[5] Hammer-Jensen argued that *Meteorology* IV was insufficiently teleological to be by Aristotle, and that its frequent appeal to "pores" (*poroi*) and "corpuscles" (*onkoi*) as explanatory agents was evidence of atomism on the part of the author. The text explains everything from the insolubility of terracotta to the combustibility of wood by reference to the presence or absence of insensible pores and the particles that can enter them. Despite the corpuscular proclivities of *Meteorology* IV, however, the current scholarly climate has changed rather decisively in favor of its Aristotelian origin, seeing in the text a development of ideas introduced in *De generatione et*

---

2. The influence of *Meteorology* IV per se is the subject of a recent dissertation by Craig Martin, "Interpretation and Utility: The Renaissance Commentary Tradition on Aristotle's *Meteorologica* IV" (Ph. D. diss., Harvard University, 2002).

3. The alchemical tradition before Paracelsus had stressed analysis, but not resynthesis. This would change with the new emphasis on *spagyria*.

4. Christoph Meinel, "In physicis futurum saeculum respicio," *Veröffentlichung der Joachim Jungius-Gesellschaft der Wissenschaften* (Göttingen: Vandenhoeck & Ruprecht, 1984); Hans Kangro, *Joachim Jungius' Experimente und Gedanken zur Begründung der Chemie als Wissenschaft* (Wiesbaden: Franz Steiner, 1968), pp. 17–34.

5. Aristotle, *Meteorology*, ed. and trans. H. D. P. Lee, pp. xiii–xxi.

corruptione and serving perhaps as a prolegomenon to Aristotle's biological works.[6]

As we have pointed out, the *Meteorology*—especially but not exclusively *Meteorology* IV—recombined with alchemy in the seventeenth century to yield a consciously experimental and corpuscularian form of Aristotelianism. The immediate precursor to this development lay in the work of Paracelsus, with its emphasis on *Scheidung*, the separation of preexistent constituents, which I discussed in chapter 2. In the early seventeenth century, Paracelsus's term *spagyria*—which originally emphasized analysis or *Scheidung* over synthesis—was subjected to linguistic analysis by the fiery polemicist Andreas Libavius, an outspoken opponent of Paracelsus who nonetheless defended chymistry. As we will see, Libavius's treatment would explicitly link *spagyria* to atomism via the intermediary of Aristotle's *Meteorology*. The facts of Libavius's life remain scanty, despite the massive tomes that he composed, but it is sure that the sanitized picture sometimes given of him as father of the modern chemical textbook does scant justice to the Saxon schoolmaster.[7] During a career that spanned positions as an academic historian and poet at Jena, as municipal physician at Rothenburg, and as director of the Gymnasium Casimirianum in Coburg, Libavius managed to become one of the premier spokesmen for the reality of chrysopoeia in early modern Europe. He defended the medieval alchemists against the pretensions of the upstart Paracelsians, whom he despised, and

---

6. Aristotle, *Meteorology* (H. D. P. Lee), pp. xiii–xxi, Furley, "Mechanics of Meteorologica IV," pp. 73–93, Louis, *Aristote: Météorologiques*, pp. xii–xv. But see Strohm, "Beobachtungen," pp. 94–115. Strohm considers *Meteors IV* to be a "Bearbeitung." A fairly recent *status quaestionis* may be found in Carmela Baffioni, *Il IV Libro dei "Meteorologica" di Aristotele* (Naples: C.N.R., Centro di studio del pensiero antico, 1981), pp. 34–44.

7. I refer primarily to Owen Hannaway, *The Chemists and the Word: The Didactic Origins of Chemistry* (Baltimore: Johns Hopkins University Press, 1975), and idem, "Laboratory Design and the Aim of Science: Andreas Libavius versus Tycho Brahe," *Isis* 77 (1986): 585–610. It is true, of course, that Libavius had a heavy influence on the seventeenth-century textbook writers in the tradition of Jean Beguin and his *Tyrocinium chymicum*. But even Beguin was openly interested in chrysopoeia. One cannot simply ignore the chrysopoetic content of seventeenth-century chymical books and pluck out the elements that seem appealing from the perspective of modern chemistry. For a critique of the excessively progressivist picture of Libavius given by Hannaway, see William R. Newman, "Alchemical Symbolism and Concealment: The Chemical House of Libavius," in *The Architecture of Science*, ed. Peter Galison and Emily Thompson (Cambridge, MA: MIT Press, 1999), pp. 59–77. For some comments on Beguin's interest in chrysopoeia, see William R. Newman and Lawrence M. Principe, "Alchemy vs. Chemistry: The Etymological Origins of a Historiographic Mistake," in *Early Science and Medicine* 3 (1998): 32–65, especially pp. 49–51.

viewed the *Summa perfectionis* of Geber as the classic text of metallurgical alchemy.[8]

In one of his endless arguments with the Erastian enemies of chrysopoeia (such as the Lotharingian physician Nicholas Guibert, whom we will examine in the following chapter), Libavius tries to find a respectable origin for the Paracelsian neologism *spagyria*. This exercise in word origins shows that Libavius already viewed *spagyria* in explicitly corpuscular terms in his *Alchymia* of 1606. Here Libavius provides several possible etymologies for *spagyria* but settles ultimately on a fusion of the Greek *span* (to pull apart) with *ageirein* (to put together):

> The moderns call it *spagiria* (σπαγειρίαν). Leo Suavius does not know from whence.... But most celebrated is that *synkrisis* and *diakrisis* of the old, called "coagulation" [and] "solution" by our artisans [i.e., chymists]. For the latter tear apart the structures of mixed bodies and break them up with their ingenious techniques and apparatus. Penetrating into the inner chambers of composite things, into the bedrooms and sanctuaries of their essences, they congregate and unite the homogeneous, while separating the heterogeneous. That is, in Greek *span* and *ageirein*.[9]

In this passage, Libavius clearly derives the Paracelsian term *spagyria* from *span* and *ageirein*. But Libavius also points out that these twin processes of analysis and synthesis correspond to the *diakrisis* and *synkrisis* of the ancients. To anyone with a knowledge of Greek philosophy, *diakrisis* and *synkrisis* would have been highly charged terms, for these are the very words that Aristotle uses when describing—and debunking—the atomism of Democritus and Leucippus. *Diakrisis* referred to the separation of atoms from one another, and *synkrisis* to their combination.[10] Is Libavius then giving a Democritean interpretation to Paracelsian *spagyria*?

The answer to this question is far from simple. As we will see, Libavius did have Democritean sympathies, but his Democritus was a far different

8. For Libavius's high regard of Geber, see his *Rerum chymicarum epistolica forma* (Frankfurt: Petrus Kopffius, 1595), vol. 1, pp. 112–113. See also Libavius, *Commentariorum*, part 1, pp. 83–84, in *Alchymia ... recognita, emendata, et aucta* (Frankfurt: Joannes Saurius, 1606); and Newman, "Alchemical Symbolism and Concealment," pp. 67–73.

9. Libavius, *Alchymia, Commentariorum in librum primum alchymiae partis I. Lib. I*, p. 77: "Spagirian (σπαγειρίαν) appellant recentes. Nescit Leo Suavius unde ... Sed celebratissima est illa veterum σύγκρισις καὶ διάκρισις coagulatio, solutio nostris artificibus dicta. Divellunt hi, perfringuntque compages mistorum adminiculis & instrumentis ingeniosis; & in penetralia compositarum rerum, cubiculaque & adyta essentiarum penetrantes, homogenea congregant, uniunt, & ab heterogeneis separant. Id est Graecis, σπᾶν καὶ ἀγείρειν."

10. Aristotle *De gen. et corr.* I 2 315b 7–10, 317a 13–14.

figure from the straightforward atomist whom we encounter in the history of philosophy. Additionally, the terms *diakrisis* and *synkrisis* and their variants were used widely by Aristotle to describe physical processes that he upheld himself, especially in the books of the *Meteorology*. By combining the Democritean and Aristotelian uses of *synkrisis* and *diakrisis*, Libavius was able to provide a description of material change that was sympathetic to Democritus while retaining the authority of Aristotle. In order to see how this is so, we must consider Libavius's long and bitter rebuttal of an earlier attack on chymical medicine by Jean Riolan, censor of the Paris medical faculty.[11] In the course of his diatribe, Riolan had attacked Democritus, apparently because one prop for the antiquity of alchemy had been the pseudo-Democritean *Physika kai mystika*, an alchemical work that was actually composed by a forger in the Roman Imperial period. This attack prompts Libavius to defend Democritean atomism in a quite colorful fashion:

> *You ridicule atomic corpuscles and their concretion and segregation.* But I say do tell, you jocular philosophers, why it is that when you generate man by means of a segregation from a liquor or from an evaporation of the humidity which Aristotle stated to come about in the body of the offspring, you do not see any such thing in the liquor beforehand? And when you inspect the white of an egg, why do you not see visible members? In the same fashion when even very thick things are dissolved in sharp waters they do not appear to the eyes, although they reappear once the humidity is removed. Salt cookers know this, and those who evaporate solutions of niter, halonitrum, chalcanthum, pure urine, clear wine, and innumerable such things. Indeed, meteorologists are forced to confess this when they create not only clouds and comets but even minerals of every sort from vapors and exhalations.[12]

Libavius is keen to show here that Aristotle himself employs the notion of associating and dissociating particles to explain such processes as

11. Debus, *French Paracelsians*, p. 57. Riolan's attack was entitled *Apologia pro Hippocratis et Galeni medicina* (1603).

12. Libavius, *Alchymia, Prooemium commentarii alchymiae*, p. 3: "*Atoma corpuscula ridetis, & eorum concretionem & segregationem.* Sed dicite quaeso vos ridiculi philosophi, cum hominem ex liquore generatis per segregationem, seu divaporationem illam humiditatis quam in corpore geniturae fieri statuit Aristoteles, num prius cernebatis aliquod tale & tantum in liquore? Cum inspicis ovi album, num visibilia vobis sunt membra? Ita cum aquis acribus solvuntur res maxime corpulentae, non apparent oculis: Et tamen segregato humore emergunt. Id norunt & qui muriam in salem excoquunt, qui dilutum nitri, halonitri, chalcanthi, urinas puras, vinum limpidum & innumera talia. Imo & meteorologici coguntur hoc fateri, cum ex vaporibus & exhalationibus non tantum creant nubes, & cometas verum etiam fossilia omnis generis."

human generation. It is only on this assumption, moreover, that one can explain the apparent disappearance of minerals in a solution and their unchanged reappearance when the solution is eliminated. This observation, repeated by Libavius in subsequent works and later elaborated by Sennert, would come to form one of the seventeenth century's popular if anecdotal indications of the corpuscular nature of matter.[13] Finally, Libavius makes an implicit appeal to the *Meteorology*, referring obliquely to several passages.[14] Interestingly, two of the three passages to which Libavius apparently alludes use the word *synkrisis* to describe the concentration of matter necessary to make clouds and comets (340a 24–341a 12 and 345b 31–346a 9a). In the Greek text to these passages, the participial and verbal forms of *diakrisis*, namely, *diakrinomenos* and *diakrinein*, are also found (e.g. at 345b 34, 346a 9a). Behind Libavius's allusions to human generation, eggs, salt cookers, and meteorologists, then, lurk *synkrisis* and *diakrisis*, the tools of the spagyrist.

Libavius's references to meteorological and alchemical processes as due to *synkrisis* and *diakrisis* reveal his longstanding affection for the *Meteorology*. This concern is apparent already in his *Rerum chymicarum epistolica forma* of 1595. In a letter there addressed to the anatomist Caspar Bauhin, Libavius argues that the principles of chymistry can be found in Aristotle and that chymists can even learn their techniques from the Stagirite. He then proceeds to quote from *Meteorology* II (358b16 ff), where Aristotle describes the evaporation and condensation of wine to "water" (ὕδωρ). Libavius argues that this "water" is not really the element

---

13. Libavius, *Alchymia triumphans*, pp. 154–155: "Ex invisibilibus etiam Deus scribitur mundum fecisse: et tamen ipse in principio fecit aquas & terram, atque ita omnium corporeorum prima rudimenta posuit elementa visibilia. Utrumque verum esse suo sensu potest; ex individuis sensu inaspectabilibus cogere seu condensare & coagulare (quomodo Hiob scribit se a Deo coagulatum esse) corpora, ut cum muria coquitur, vel ex aquis causticis reducuntur metalla, ex dilutis alumina, perlae coralia, & quid non resolutorum? Et deinde ex elementis seu principiis facere. Est enim alterum alteri subalternum, passioque solutionis indivisibilia communis omnibus corporibus. Peripatetici autem disputant contra Democriti atoma; Sed logice magis quam ad sensum. Corpus ullum negant esse indivisibile. At cum ad actum divisionis perventum est, ratio producitur de continuis semper divisibilibus deficiente sensu. Sapiunt certe illa atoma chymicam experientiam, qua in aquis pellucidissimis, subtilissimisque & per cola arctissima transmeantibus, cognoscuntur inesse corpora satis crassa. Reductione enim facta emergunt. Illud ipsum docet etiam metallorum secundum Aristotelem origo, & cernimus etiamnum hodie ex fumis concrescentibus coalescere corpora admodum densa & gravia; quae si resolvas non ex uno constare animadvertuntur." Sennert, *De chymicorum* (1629), p. 212: "Ita in lixivio & muria, e qua sal, nitrum, & vitriolum coquitur; in urina, in vino sal, qui inest, non conspicitur: at separato humore, facile se conspiciendum praebet."

14. Aristotle *Meteorology* 341a 9–28, 344b 8–10, 378a 15–31.

that goes by that name but rather a "mixt," which is "hot, burning, and flavorful" (modern ethyl alcohol). In the same vein, Libavius uses the *Meteorology* to explain the fact that aquosity can be used to elevate the dry components of a substance, or in laboratory terms, that it is usually easier to distill a material that has been dissolved in liquids than to elevate it while dry. This is because the elevated moist vapor in the flask "gathers up" (συμπεριλαμβάνειν) the dry particles just as Aristotle says that the south wind, beginning as a cold breeze, gradually absorbs large quantities of hot exhalation (ἀναθυμίασις) from the places through which it passes (358a 33–35).[15] In these passages, then, we see Libavius focusing on two characteristics of the *Meteorology* that were particularly attractive to him—a preference for explaining the decomposition of substances into "mixts" such as alcohol rather than into pure elements, and the treatment of these mixts in terms of small particles or corpuscles. The same characteristics emerge again in the remaining part of the letter.

Libavius's letter to Bauhin then progresses to the famous example from *Meteorology* II of seawater supposedly filtered through a wax container (358b 34 ff.). According to Aristotle, the wax jar will be found to contain drops of fresh water after it has remained in the ocean for a time. As he says (in Lee's translation): "[T]he earthy substance whose admixture caused the saltness is separated off as though by a filter." After considering this passage, Libavius points out that Aristotle's chymical knowledge is further revealed when *Meteorology* II describes the production of alkali by means of a lixiviation from the cinders of rushes and reeds (at 359b 1–4). According to Libavius, the process works because the smallest particles (*tenuissimae partes*) of the salt, separated from the plant cinders by the fire, are retained in the water of solution. When the water is then removed by evaporation, the purified salt remains. Finally, Libavius passes to a passage from *Meteorology* IV that is perhaps related to the section from his *Alchymia* quoted above, where the evaporation of water from brine by salt cookers was used as a support for Democritus. At *Meteorology* IV 384a 4–7, Aristotle says that must, and some other types of wine, solidify if they are boiled—"in all such cases it is the water that is driven off in the process of drying" (quoting Lee's translation). Libavius expands on this passage, saying that Aristotle uses the term "must" (γλεῦκος) to mean all sorts of "extracted fluids" (*succos extractos*), and that these all coagulate when their water is removed, as in the case of evaporation. As Libavius says, "Experience shows that [juices] do not

15. Libavius, *Rerum chymicarum, Liber primus*, pp. 248–255.

sink into themselves (*in se considant*), but rather their water departs."[16] The point of this seems to be that evaporative processes do not involve a transmutation, as of water into earth or air, but instead a mere removal of the volatile particles making up the fluid from the fixed corpuscles that are left behind—in other words, a *diakrisis*.

We see then that even in the 1590s Libavius was deeply interested in the *Meteorology*, including book IV. From that work he was able to derive both corpuscular explanations of phenomena and references to decomposition products that were still mixts—and therefore partook of all the specific properties that we encounter in the physical world—rather than being simple elements.[17] Yet we will see that Libavius was far from being a genuine atomist or even a systematic corpuscularian. Already in his *Rerum chymicarum epistolica forma*, Libavius expresses his "Democritean" sympathies, saying "since Democritus did not gaze upon the external face of things, but rather the internal nature just as he learned from chymistry, he laughs at human deeds and holds them to be stupid. This is that light of genius and wisdom that the true *spagyrus* obtains."[18] This passage contains no explicit reference to atoms. It alludes, rather, to the tradition of Democritus as an anatomist presented in the pseudonymous *Epistola ad Damagetum*, a letter supposedly written by Hippocrates but actually forming part of an epistolary novel composed by an anonymous forger around 40 B.C. As Christoph Lüthy has recently shown, this letter forms one of the major sources of a complex "Democritean" tradition that was still strong in the seventeenth century.[19] In the *Epistola ad Damagetum*, the famous physician of Kos says that at the request of

16. Libavius, *Rerum chymicarum, Liber primus*, p. 252: "Quod autem non in se considant, sed aqua revera egrediatur arguit experientia."

17. The latter point is further emphasized in Libavius, *Alchymia, Prooemium commentarii alchymiae*, on p. 63: "In 4. *meteor.* Aristoteles non dicit resolvi omnia in elementa; sed sublato termino per calorem ambientis in putredine humidi, & calorem insitum expirare, subeunte humido alieno & frigido, quibus materiae simul designantur. Chymici illas expirationes studiose colligunt, & Aristotelis κόπρον cum aliena humiditate elementis accensent; collectas expirationes essentiae nomine vocant, & in hac inveniunt totam vim rei. Num usque adeo ab Aristotele tibi videntur alieni?"

18. Libavius, *Rerum chymicarum, Liber primus*, p. 157: "Unde cum Democritus non externum rerum vultum contemplaretur, sed naturam internam perinde ut ex chymia didicerat; ridet humana facta & pro stolidis habet. Haec est illa ingenii & sapientiae lux quam lucratur spagirus verus."

19. Christoph Lüthy, "The Fourfold Democritus on the Stage of Early Modern Science," *Isis* 91 (2000): 443–479. For the *Epistola ad Damagetum*, see Thomas Rütten, *Demokrit, lachender Philosoph und sanguinischer Melancholiker: Eine pseudohippokratische Geschichte* (Leiden: Brill, 1992), passim.

the citizens of Abdera he paid a visit to Democritus, who had gone into seclusion and was suspected of having lost his mind. Hippocrates found him sitting on a stone, scribbling furiously, and surrounded by books and dissected animals. When one of the Abderites began lamenting this madness, Democritus burst into laughter. Hippocrates then approached him alone, whereupon the marvelously lucid Democritus explained that he was writing a book on madness and hoped to learn the site of black bile—the cause of insanity—from his dissections.[20]

As we saw above, however, Libavius does argue for the existence of Democritean "atoms" in the 1606 *Alchymia*. Yet we must be very careful in ascribing a coherent corpuscular matter theory to the Saxon chymist. Libavius is notorious for his inconsistency, a fact that Daniel Sennert was already complaining of in 1619.[21] In addition, Libavius was deeply enamored of the esoteric, despite modern attempts to portray him as a champion of open communication.[22] He was a seasoned interpreter of alchemical symbolism, expending a considerable portion of his gargantuan energy unraveling the secrets of John Dee's *monas hieroglyphica* and other riddling *figurae*. Since Libavius himself accepted the genuineness of the alchemical *Physika kai mystika* attributed to Democritus, it should not surprise us that he read even the accounts of the Abderite's atomism as containing veiled secrets.[23] Hence Libavius refers to the "symbolic philosophy of Democritus" (*Philosophia Symbolica Democriti*).[24] By hiding his real meaning, the Libavian Democritus managed to avoid the fate of the Stoics, whose views had been "exploded" as a result of their openness. In this fashion, Libavius could explain away unpleasant features of the Democritean philosophy such as the Abderite's belief in an infinite number of worlds. Since Democritus believed in the doctrine of the macrocosm and microcosm and thought that the beings that undergo generation and corruption—such as men, beast, plants, and minerals—were all microcosms, he was only stating that these beings are innumerable, not that the universe is literally infinite.

20. Friedel Pick, *Joh. Jessenius de Magna Jessen, Arzt und Rektor in Wittenberg und Prag hingerichtet am 21. Juni 1621*, in *Studien zur Geschichte der Medizin* (Leipzig: Ambrosius Barth, 1926), vol. 15, pp. 122–123. A much deeper treatment is found in Rütten, *Demokrit*.

21. Sennert, *De chymicorum* (1619), p. 301.

22. See Newman, "Alchemical Symbolism," pp. 59–77, where Libavius's interpretation of John Dee's esoteric *monas hieroglyphica* is described at length.

23. Libavius, in *Alchymia, Prooemium commentarii alchymiae*, p. 2, where Libavius refers to the "de lapide Philosophorum libellus" of Democritus.

24. Libavius, *Alchymia, Prooemium commentarii alchymiae*, p. 3.

Libavius has similar things to say about Democritean atomism in the *Alchymia*. If Democritus were alive today, Libavius insists, he would admit the existence of Aristotle's prime matter, at least "after a certain fashion" (*quodammodo*). In the examples of salts and other solids invisibly dissolved in liquids, Democritus would have no problem with the claim that this dispersed *prima materia*, once collected together, would then be "mutated and [its] acts educed from potency" (*mutari & ex potentia educi actus*). Libavius develops this supposed harmony between the Stagirite and the Abderite still further in his *Alchymia triumphans* of the following year (1607). Bluntly stating that "Democritus does not disagree with Aristotle" (*Non dissentit Democritus ab Aristotele*), Libavius explains that Democritus did not deny the Aristotelian doctrine of *mixis*. To the contrary, each Democritean atom is a perfect mixt, and the Abderite may even have thought that his atoms were created as such by God. What is more, Libavius suggests that Democritus thought that his atoms did not of themselves supply the gross matter of the physical world, but that macrolevel bodies had to be built up out of a combination of "atoms" and the four Aristotelian elements.[25] Here we are far indeed from the ontological economy of the genuine Democritus with his atoms of matter differentiated only by size, shape, and position, separated by pure void (the πλῆρες and κενόν of *Metaphysics* A 985b5). Nonetheless, Libavius asserts, it would be fatuous to deny that many processes occur as a result of the association and dissociation of these divinely mixed "atoms" (in combination with the four elements). This is the case, for example, when particles of butter, whey, and curds separate out of milk. In an apparent reference to the formation of the human fetus that he had already considered a case of *synkrisis* in the *Alchymia* of 1606, Libavius even says that "we are coagulated by God like cheese" (*Ut caseus coagulati a Deo sumus*).[26]

From our analysis up to this point, it might appear that Libavius did achieve a straightforward synthesis of the doctrines in Aristotle's *Meteorology* with his understanding of Democritean atomism. Even the example of the separation of milk into curds, butter, and whey is found in *Meteorology* IV (384a 20–25), where Aristotle explains this as resulting from the coagulation (σύστασις) and separation (χώρισις) of these ingredients. To Libavius, this process clearly involves the association and

---

25. Libavius, *Alchymia triumphans*, pp. 160–161: "Quid si diceret Democritus sua atoma pridem esse a Deo mista, nec indigere concretione inde usque ab ovo, quanquam porro coalescere cum elementis in corpus robustius & crassius possint?"

26. Libavius, *Alchymia triumphans*, p. 161.

dissociation of particles of curds, butter, and whey. This sounds at first like a quite coherent corpuscular theory, but other Libavian comments suggest that things are not so simple. In his *De mundi Corporumque mixtorum elementis* (On the Elements of the World and of Mixed Bodies), a disputation over which Libavius presided at the Gymnasium Casimirianum in 1608, he distinguishes between the four elements, which act like "wrappings" (*involucra*) or "shells" (*cortices*) and the secret "first principles" (*principia prima*) that lie hidden within them.[27] Later in the same text, he argues that the four elements really do undergo Aristotelian *mixis* but distinguishes these "vulgar" elements from those of the *Hermetici philosophi* (Hermetic philosophers).[28] The *Hermetici*, by whom Libavius clearly means the alchemists, view the vulgar elements, once again, as mere "boxes" (*arcae*) or "repositories" (*repositoria*), which hide the real essences of things. The vulgar elements conceive a *semen* (seed) transmitted by the heavens; this seed then uses the elements as a womb in which it can mature. Libavius continues to say that "all mixts are therefore essentially from the heavens; but in order to be sensible, they bear the bodies of the elements with themselves, from which they can never be totally separated."[29] The claim made here that all mixts receive their essence from the heavens should alert us to the fact that Libavius is talking about the very entities that he earlier identified as Democritean atoms, for in the *Alchymia triumphans* of 1607 he had already suggested that Democritus believed in a creator-god who formed the atoms as perfect mixts. And there too he argued that these divine atoms had to combine with the grosser elements of Aristotle in order to produce sensible bodies. These ideas are elaborated considerably in the *De mundi*:

> Aristotle otherwise often laughs at Anaxagoras, Empedocles, and the like on the topic of the elements and on their combining with similars. In book 3 of *De caelo*, chapter 7, text 56 [305b1–5], he has imputed apparent [i.e., illusory] generation to Empedocles and Democritus, as though it came about in a vessel by means of segregation. But the *Hermetici* deny that these philosophers have been understood by Aristotle, as they [Empedocles and Democritus] did not assent to that sensible and unsophisticated separation; rather, they believed in an essential generation from roots and *seminia* [i.e., *semina* or "seeds"] of things that God sealed up in the elements and conjoined with them.[30]

27. Libavius, *De mundi Corporumque mixtorum elementis*, [A3v].
28. Libavius, *De mundi*, [B1r–v].
29. Libavius, *De mundi*, [B4v].
30. Libavius, *De mundi*, [C1r]: "Aristoteles alioquin saepe irridet Anaxagoram, Empedoclem & similes de Elementis, & eorum ad similaria concursu. In 3. De coelo cap. 7. Tex. 56. Tribuit

In the above passage, Libavius makes it quite clear that the chymists (*Hermetici*), with whom he identifies, have extracted a secret meaning from the *Philosophia Symbolica Democriti*. In effect, they have turned Democritus into a Renaissance Neoplatonist, of the stamp of Marsilio Ficino and Agrippa von Nettesheim.[31] Lurking behind Libavius's *Hermetici* one may also discern the erudite Danish systematizer of Paracelsus, Petrus Severinus, whose remarkably influential *Idea medicinae* of 1571 had placed great stress on the action of *semina* on matter. Severinus's *semina*, which owed a debt both to the "seminal reasons" (*rationes seminales*) of Saint Augustine and to the hylozoism of a host of Renaissance thinkers, were dimensionless centers of activity that governed the development and action of physical bodies. They found their material homes within the elemental realm but were not coextensive with the four elements, being intermediary between the divine and physical worlds. Despite their "metaphysical" status, Severinus's *semina* were far from being incompatible with atomistic or corpusuclar theories of matter. Robert Boyle, among others, would employ "seminal principles" derived in part from Severinus as a means of explaining the interactions of otherwise brute corpuscular matter.[32] Hence it is no surprise to find Libavius equating *semina* and Democritean atoms at the inception of the seventeenth century. From the perspective of Libavius's *De mundi*, those genuine mixts that supply the essential being to all things are produced by God as *semina*. Once they are sealed up in elementary matter, they act on it to produce the myriad generations and corruptions of the physical world. At the same time, the elemental *mixis* of the Aristotelians also

---

Empedocli, & Democrito apparentem generationem, quasi per segregationem tanquam ex vase fiat. Sed Hermetici negant istos Philosophos intelligi ab Aristotele, utpote non sensibilem, & rusticam illam secretionem innuentes, sed generationem essentialem ex radicibus, & seminiis rerum, quae DEUS Elementis inclusit, & cum ipsis conjunxit." The term *semina* seems to be used here as a synonym for *semina*, although *seminium* should really mean either the action of procreation or a "breed." See the *Oxford Latin Dictionary*, s.v. "semina. "

31. For the influence of Agrippa on early modern alchemy, see Newman, *Gehennical Fire*, pp. 209–227.

32. For the compatibility of Severinus's *semina* and atoms or corpuscles, see the important article by Jole Shackelford, "Seeds with a Mechanical Purpose: Severinus' Semina and Seventeenth-Century Matter Theory," in *Reading the Book of Nature: The Other Side of the Scientific Revolution*, ed. Allen G. Debus and Michael T. Walton (Kirksville: Sixteenth-Century Journal Publishers, 1998), pp. 15–44. My treatment of Severinus owes much to Shackelford, *A Philosophical Path for Paracelsian Medicine: The Ideas, Intellectual Context, and Influence of Petrus Severinus, 1540–1602* (Copenhagen: Museum Tusculanum Press, 2004), especially pp. 160–185. See also Hiroshi Hirai, "Le concept de semence dans les théories de la matière à la Renaissance" (Ph.D. diss, Université Lille III, 1999), vol. 1, pp. 162–200.

occurs, and it should not be opposed to the mixture of the *Hermetici* as though the two contradicted one another.[33] But this Aristotelian *mixis* is of little importance, for the primary qualities—heat, frigidity, humidity, and aridity—produce only hot, cold, wet, and dry things and the elements only produce other elements.[34] It is the *semina* hidden within the elements that determine essential characteristics and supply particular qualities to physical things. These seedlike particles do not of course undergo mixture themselves, for they are the perfect mixts that God himself has created in the beginning as principles of specificity.

In this fashion, then, Libavius upholds both the reality of Aristotelian *mixis* and the existence of divinely mixed "atoms." In his view, the production of perfect mixts does indeed occur in the sublunar realm, and at times Libavius even explicitly denies that the ingredients of such mixtures continue to exist *in actu* once the process of *mixis* has taken place.[35] Clearly the acceptance of such total mixture among the elements eliminates Libavius from the tribe of either genuine atomists or consistent corpuscularians. At the same time, however, he asserts that this "vulgar" sort of mixture does not occur among his "atoms," which he equates with the *semina* of the *Hermetici*. The "atoms" (*semina*), having been perfectly mixed by God at their creation, undergo no further *mixis*. The goal of chymistry, in his view, is precisely the separation of these *semina* from the crude impurity supplied by the four vulgar elements. This chymical separation, performed by such processes as calcination, sublimation, and above all, distillation, is clearly viewed by Libavius as an Aristotelian *diakrisis* of the type described in the *Meteorology*. To summarize the position of Libavius, then, the Saxon chymist does indeed utilize *synkrisis* and *diakrisis* as explanatory tools and even thinks of them in terms of the aggregation and separation of corpuscles. Yet he does not commit himself to the continued existence of those corpuscles within a body once it has been composed. In scholastic terminology, Libavius does not commit himself to the reality of *minima inexistentia* (minima existing within) except for the discrete *semina* that lurk within the depths of gross elemental matter.

The endless qualifications and digressions in Libavius's matter theory reflect, in part, his attempt to combine sources of such extreme diversity as ancient atomism, Aristotelian hylomorphism, medieval alchemy,

---

33. Libavius, *De mundi*, [D4v].
34. Libavius, *De mundi*, [Cr], [E2v].
35. Libavius, *De mundi*, [D2v]; see also [B1r].

Renaissance Neoplatonism, and the *spagyria* of the Paracelsians. In addition, Libavius usually employed a pugnacious style of rebutting his opponents' views rather than building up an easily intelligible system of his own. Nonetheless, he exercised a powerful impact in the first half of the seventeenth century, becoming the founder of an influential movement to write chymical textbooks. Some of these, such as the hugely popular *Tyrocinium chemicum* of Jean Beguin, retained the traces of Libavius's corpuscular explanation of Paracelsian *spagyria*. The 1612 edition of Beguin's *Tyrocinium* provided a number of definitions of chymistry, one of which defined the cognomen *spagyria* as follows: "If [one] should call it *spagyria*, he denotes its principal operations, namely *synkrisis* and *diakrisis*."[36] Here and elsewhere we see the results of the Libavian use of Aristotle's meteorology to elide the boundaries between peripatetic *mixis* and atomism. Driven by Erastus and his followers, who had built their rebuttal of alchemy on a strident defense of Aristotelian mixture theory and hylomorphism, Libavius responded by emphasizing the corpuscularian features within Aristotle's *Meteorology*. At the same time, by invoking the late antique tradition of Democritus the alchemist, Libavius was able to present the chrysopoetic art as a discipline that predated Aristotle. The crowning edifice to this imaginative historical exercise lay in Libavius's claim that Democritus himself was a proponent of Aristotelian perfect mixture and that salient features of the Abderite's alchemical atomism could still be found buried in the matter theory of the Stagirite, like so many *semina* hidden within their elementary "boxes." Libavius's esoteric reading of Aristotle allowed him to conclude as a peripatetic that "God did not create [the world] from elements alone, but from a celestial seedbed sealed up in the elements."[37]

As the seventeenth century progressed, Libavius's strategy of merging Aristotle and Democritus would contribute strikingly to the movement to reintegrate Geberian corpuscular theory with the particulate ruminations of Aristotelian meteorology. We will soon see the fruits of this movement when we examine the influential natural philosophy of Daniel

36. Jean Beguin, *Tyrocinium chymicum recognitum et auctum* (Paris: Matheus le Maistre, 1612), pp. 1–2. For Beguin, see T. S. Patterson, "Jean Beguin and His Tyrocinium Chymicum," *Annals of Science* 2 (1937): 243–298. See also William R. Newman and Lawrence M. Principe, "Alchemy vs. Chemistry: The Etymological Origins of a Historiographic Mistake," *Early Science and Medicine* 3 (1998): 32–65; especially 49–51.

37. Libavius, *Alchymia, Prooemium commentarii alchymiae*, p. 20: "Nimirum non ex elementis solum Deus res procreavit, sed omnino ex coelesti simul incluso elementis seminario." In Libavius's text, this passage is followed by an elaborate justification drawn from Aristotle, *Meteorology* I and *De generatione animalium* I–II.

Sennert. As for the Libavian importation of celestial *semina* into matter, while this was alien to the spirit of Geber, it allowed later chymists to explain the activity and qualitative characteristics of particular materials in a way that the *Summa perfectionis* had not attempted to do. Working from analytical laboratory operations whenever possible, the *Summa* had for the most part eschewed discussion of qualities and observations that could not be explained by means of the author's corpuscular theory. For example, Geber's explanation of the difference in color between gold and silver is based on the assumption that the nobler metal contains a red sulfur while the less precious one is colored by a white sulfur. But why is it that these sulfurs themselves have different colors? No attempt at an explanation is given. As for the fact that sulfur is inflammable whereas mercury is not, Geber's explanation consists of the claim that sulfur is fiery, whereas quicksilver is watery. Although his corpuscular theory allowed Geber to explain why some sulfur is more prone to combustion than other sulfur on the assumption that the former was made up of smaller particles, Geber was unable to provide a corpuscular explanation that would reveal why sulfur per se was inflammable and mercury was not.

For the most part, the *Summa* merely took the chemical qualities and specificity of the principles sulfur and mercury as givens: the author then used the principles to account for the generation of the different metals without descending further into the underlying causes of chemical variation in the principles themselves. Although his constant appeal to varying particle size and homogeneity allowed Geber to explain differences in volatility and fixity, specific gravity, and resistance to the destructive power of fire and heat, and his concept of "very strong composition" could account in principle for bonding between corpuscles, Geber made no sustained attempt to explain other phenomena such as particular chemical affinities or repulsions, not to mention the tastes, smells, and colors of individual chemicals. In addition, while he could account for the greater or lesser combustibility of a given sample of sulfur on the basis of its having smaller or larger particles, he gave no corpuscular explanation for the complete absence of combustibility in other powdery materials, such as the calces of the metals. Chymists such as Severinus, who adopted a purely qualitative approach to their discipline based on seminal principles, no doubt saw themselves as avoiding this weakness in Geberian alchemy when they provided a factotum that could account for any type of chemical specificity, attraction, or repulsion. At the same time, however, they ran the risk of losing the advantages in theoretical economy and observational fit offered by Geber's corpuscular explanations. Libavius's association of seminal principles with atoms had the advantage of opening a new path by explicitly combining the two approaches. Now one could employ the structural explanations offered by Geberian alchemy while also providing an account of qualitative differences that evaded a simple corpuscular description.

But this was not the end of the story. Libavius's conversion of Democritean atoms to perfect Aristotelian mixts provided his successors with the way to yet another opening. Why not carry the Aristotelianizing process inaugurated by Libavius a step forward by treating the dimensionless *semina* of the *Hermetici* as peripatetic substantial forms? One could then have all the advantages of the Geberian corpuscular tradition—now clothed in the explicit vesture of Aristotelian meteorology—while also providing a convincing Aristotelian account of the features that Geber left largely unexplained: affinity, repulsion, corrosiveness, causticity, and the myriad other specific qualities of reagents whose number grew daily as the art of chymistry expanded its purview and repertoire. Nowhere would more eager proponents of this tactic emerge than within the eclectic but self-consciously peripatetic atmosphere of the University of Wittenberg.

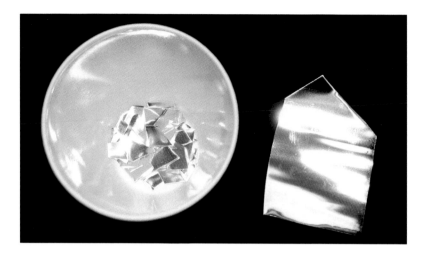

FIGURE 1. A thin sheet of technical grade silver partially cut into small pieces. This is the starting point of one of Daniel Sennert's most famous "reductions to the pristine state." The experiment was subsequently adopted by Robert Boyle. The following sequence of processes was carried out in the Indiana University Chemistry Department in the laboratory of Cathrine Reck. Photograph by the author.

FIGURE 2. The silver pieces undergoing dissolution in nitric acid (aqua fortis). The blue color comes from the copper commonly found in technical grade silver. Technical silver was employed intentionally to approximate the materials available to Sennert. Many premodern writers remark on the blue color found in silver salts and solutions. The brownish cloud rising above the solution is nitrogen dioxide gas, produced during the dissolution of the metal.

FIGURE 3. A completely clear, blue solution of the technical silver in nitric acid. The solution can be filtered without leaving a residue, as Sennert and Boyle point out.

FIGURE 4. A solution of potassium carbonate (salt of tartar) being poured into the solution of silver and nitric acid in order to initiate precipitation. More nitrogen dioxide is perceptible above the solution.

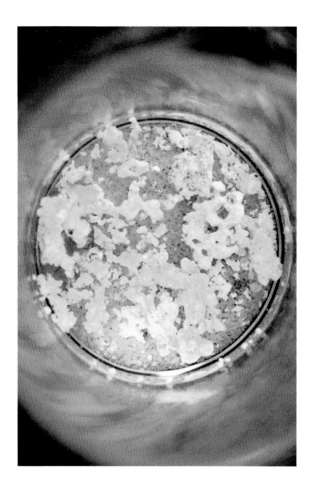

FIGURE 5. The precipitate, a curdled mass of mostly yellow silver carbonate forming at the surface of the silver-nitric acid solution after sufficient potassium carbonate is added. The precipitate eventually sinks to the bottom of the beaker.

FIGURE 6. The silver carbonate in a crucible after being filtered and washed.

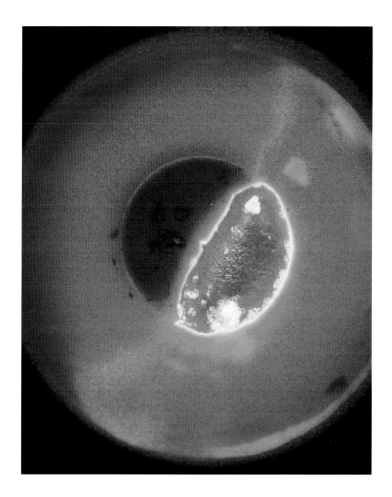

FIGURE 7. Hot silver that has been freshly reduced in a crucible from the silver carbonate precipitate by heating at about 1000° C. Simple heating of the silver carbonate yields metallic silver, carbon dioxide, and oxygen.

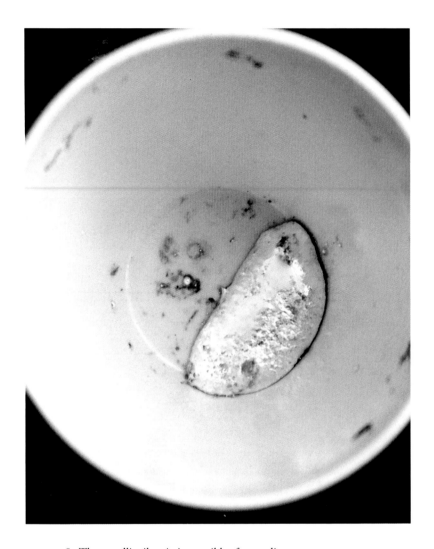

FIGURE 8. The metallic silver in its crucible after cooling.

TWO

Daniel Sennert's Atomism and the Reform of Aristotelian Matter Theory

4

The Corpuscular Theory of
Daniel Sennert and Its Sources

The "atomism" of Andreas Libavius was a rather evasive hybrid of alchemy, Renaissance Neoplatonism, and Aristotelian mixture theory, onto which the Saxon schoolmaster grafted his own esoteric interpretation of Democritus. The loose alliance that Libavius proposed between Paracelsian *spagyria* and Democritean atomism was based both on the analysis of gross matter into corpuscles by means of "chymical" techniques such as calcination and distillation and on Libavius's assimilation of atoms to unimaginably tiny *semina*. As we have seen, the analytical part of this approach had already been pioneered by Geber and other alchemists, although without either the hylozoic *semina* or the explicit allegiance to Democritean *synkrisis* and *diakrisis* that Libavius had yoked to *spagyria* in the form of analysis and synthesis.

It is now time to pass from the vaguely "atomist" intuitions of Libavius to a more decisive figure in the history of matter theory, namely, the Wittenberg professor of medicine Daniel Sennert. The fact that Sennert was an influential proponent of atomism in the seventeenth century has been well known since the appearance of Kurd Lasswitz's famous history of atomism, *Geschichte der Atomistik*, published in 1890.[1] Yet the

---

1. It must be noted at the outset that Sennert uses the term "atom" to mean a corpuscle that resists division by means of laboratory techniques, in the same way that Geber speaks of "partes." Sennert's "atoms" would not qualify as Democritean indivisibles according to our modern

trajectory by which Sennert passed from a belief in the purely homogeneous mixture of Aristotle to a rather outspoken atomism remains something of a mystery. The existing scholarship has put relatively little energy into addressing either the development of corpuscularian notions in Sennert's thought or his motivations in becoming a self-styled atomist. Several scholars who have made the first tentative steps in this endeavor have attempted to locate Sennert's atomistic inclinations in his Lutheran religiosity or in his concerns as a professor of medicine at Wittenberg.[2] While neither of these factors is to be discounted, a close look at the chronology of Sennert's developing atomism points directly to his gradually increasing appreciation of chymistry as the prime mover in his shift from the commonplace early modern view of Aristotle's matter theory to the atomistic Aristotle of the *De chymicorum* and Sennert's later work.

In the interest of addressing this issue, we will focus mainly on Sennert's early publications, leading up to and including his first atomist manifesto in the *De chymicorum* of 1619. In the course of this analysis, we will see that Sennert is a primary representative of the fusion between Aristotelianism and alchemy that we have already discussed at some length. We will also find that this amalgamation was mediated by

---

knowledge of the Abderite's philosophy. But as we shall see, Sennert interpreted Democritean atomism in a way that took the emphasis off of absolute indivisibility, either physical or mathematical.

2. See, for example, Michael Stolberg, "Particles of the Soul: The Medical and Lutheran Context of Daniel Sennert's Atomism," in *Medicina nei secoli* 15 (2003): 177–203. Stolberg's article, while giving a provocative account of Sennert's theologically tinted ideas on human generation, is explicitly predicated on the problematic belief that (pp. 177–178) "the empirical basis of early modern atomism was in fact rather shaky and its explanatory powers were limited in comparison to conventional Aristotelian physics." This claim gives Stolberg the justification for asserting that Sennert's turn to atomism must be placed in some other quarter than arguments based on empirical evidence, such as the German academic's Lutheranism. Stolberg's claim, of course, runs counter to the theme of the present book, which attempts to show exactly why Sennert's use of the reduction to the pristine state was compelling experimental evidence for the correctness of his atomistic theory both to him and in the eyes of other seventeenth-century figures. Stolberg depends here on Christoph Meinel, "Early Seventeenth-Century Atomism: Theory, Epistemology, and the Insufficiency of Experiment," *Isis* 79 (1988): 68–103. But Meinel's introductory claim (p. 68) that seventeenth-century atomism's "empirical background was weak, and not one of its alleged proofs would be accepted by today's scientific standards" by no means justifies Stolberg's suggestion that extrascientific motives were the real prime movers in Sennert's atomism, since it would be anachronistic to hold Sennert up to the standards of today's cyclotrons, cloud chambers, and X-ray diffraction photography. There are also chronological problems with Stolberg's perspective, since he is forced to rely on Sennert's late (1636) *Hypomnemata physica* for the explicit linkage of Sennert's atomism with his theology, whereas he had become a self-described atomist by 1619.

Libavius, although Sennert would work great changes on the perplexing matter theory of the irascible schoolmaster.

In 1600, the young Sennert, who had received his master's degree at Wittenberg only two years earlier, published the final installment of what would later be expanded to become one of his most popular works—the *Epitome naturalis scientiae*.[3] This early version of the *Epitome* consists of twenty-six disputations written by Sennert on the subject of natural philosophy and upheld by his students in 1599–1600.[4] The disputations were printed separately as pamphlets, followed by a title page and dedications, also printed in pamphlet form. There is no appeal to Democritean atomism in this proto-*Epitome*. To the contrary, Sennert presents himself as a staunch defender of Aristotelian philosophy and accepts Aristotle's well-known depiction and refutation of Democritus as a point atomist.[5] Sennert's adherence to peripatetic philosophy is made quite explicit even in the preface to his disputations:

> I have gathered together the quite probable and select views of Aristotle and of his most outstanding interpreters on natural philosophy (for how could I have narrated all in a short time and a few pages?), and I have collected them into a few chapters, and unless a signal necessity forced me, I have not deviated from them (which, however, I hope has happened very rarely).[6]

A close examination of the 1600 *Epitome* reveals that this is no exaggeration. Sennert refers to Aristotle as "the eagle of the philosophers, whom we will follow" and recounts an anecdote from the Renaissance scholar Caelius Rhodiginus in which Plato is supposed to have left a discussion among philosophers because Aristotle, "the philosopher of truth," had

---

3. Christoph Lüthy and William R. Newman, "Daniel Sennert's Earliest Writings (1599/1600) and Their Debt to Giordano Bruno," *Bruniana et Campanelliana* 6 (2000): 261–279.

4. Daniel Sennert, *Epitome naturalis scientiae, comprehensa disputationibus viginti sex* (Wittenberg: Simon Gronenberg, 1600).

5. Sennert, *Epitome* (1600), *Disputatio V, Thesis XII:* "An scilicet ex indivisibilibus continuum componatur? An vero ex divisibilibus? &, an continuum sit divisibile in infinitum? Fuerunt olim in ea opinione, quae statuit, continuum ex indivisibilibus componi, Pythagoras, Democritus, Leucippus, & plerique alii.... Sed illorum refutationem Aristoteles suscepit 6. Physicorum... In utramque igitur partem etsi non leves sunt rationes, nos tamen ex Aristotelis partibus stabimus, his adducti rationibus."

6. Sennert, *Epitome* (1600), preface: [A2v] "Aristotelis enim, atque praestantissimorum ejus interpretum maxime probabiles & selectas de rebus naturalibus praecipuis (omnia enim parvo tempore, paucisque pagellis comprehendere qui potuissem?) collegi sententias, easque in pauca redegi capita, nec ab iis, nisi insignis urgeret necessitas, (quod tamen rarissime factum spero) secessi."

not arrived, and therefore "intellect" was absent.[7] In accordance with this view, the earnest young Sennert does his best to convey the marrow of peripatetic philosophy as gleaned from the most up-to-date scholastic authors, such as Jacobus Zabarella, Francisus Toletus, and Benedictus Pereira.[8]

While stressing his allegiance to Aristotle, the Sennert of 1600 is demonstrably cool towards two topics that would later be associated with his name—namely, chymistry and atomism. His early reserve towards alchemy is not surprising, perhaps, since Pereira, one of his favorite sources, had treated the subject at some length in a rather deprecating vein.[9] At any rate, Sennert says the following in the two final theses of his fourth disputation, *De natura*.

> Then we ask—Can art produce the works of nature? We respond that art cannot by its own power make the works of nature, but [it can do so] by applying natural agents to patients. From this question arises a third—can alchemy, by applying natural agents to patients, make true gold? Many of the more recent authors affirm the possibility of this, even if it has not actually been discovered. But we doubt whether their arguments show this, for many reasons. [10]

It may seem incongruous to modern readers that Sennert would use alchemy to solve the question of man's ability to imitate nature, but this was standard practice among late medieval and early modern scholastics. The *quaestio famosa* whether man can really make natural products was usually solved by coming down for or against the manufacture of artificial

7. Sennert, *Epitome* (1600), *Disputatio II, Thesis XXXVI.*
8. Jacobus Zabarella (1533–1589) had a long and illustrious career as a professor of logic and of natural philosophy at the University of Padua. Franciscus Toletus (1532–1596) was a celebrated Jesuit philosopher, teaching for a number of years at the Collegium Romanum and eventually becoming a cardinal. Benedictus Pereira (1535–1610) was also a famous Jesuit author; he taught philosophy and theology at the Collegium Romanum for many years. For details on these authors and their works, see Charles Lohr, *Latin Aristotle Commentaries* (Florence: Olschki, 1988), s.v. "Zabarella, Jacobus," "Toletus, Franciscus," and "Pereira, Benedictus."
9. Sylvain Matton, "Les théologiens de la Compagnie de Jésus et l'alchimie," in *Aspects de la tradition alchimique au XVIIe siècle,* ed. Frank Greiner (Paris: S.É.H.A., 1998), pp. 383–501; see pp. 432–438.
10. Sennert, *Epitome* (1600), *Disputatio IV, Theses XL and XLI:* "Deinde quaerimus: An ars possit efficere opera Naturae? Respondemus: Propria vi Artem opera Naturae efficere non posse; sed applicando Naturalia agentia patientibus. XLI. Ex hac quastione oritur Tèrtia: An Alchymia, applicando naturalia agentia patientibus, possit verum aurum efficere. Possibile hoc esse, etsi nondum sit compertum, Recentiores multi affirmant: An tamen rationes eorum id satis evincant, multis modis dubitamus."

gold.[11] Now if we turn our gaze from the 1600 proto-*Epitome* to the 1618 *editio princeps* of the *Epitome* in book form, we shall see that despite the great similarity of the early pamphlets to the book, Sennert's position on alchemy has utterly changed. In the 1618 *Epitome*, instead of denying the reality of chymical transmutation, Sennert says that he agrees with Aristotle that it is a weakness of the mind to seek reasons for a thing to be such-and-such when we know from the facts that the thing is otherwise. And since there are vitriol fountains in Smolnitz and Goslar where iron is transmuted into copper, the facts are clear—the transmutation of metals is a demonstrated reality. "Nor do natural waters alone perform this," he continues, "but the same can also be done by art."[12] While the Sennert of 1600 had dismissed the possibility of transmutation as so unlikely that it was not worth arguing the point, the Sennert of 1618 dismissed all argument against transmutation as being sheer sophistry in the face of the facts! Clearly something radical had happened between 1600 and 1618—in short, Sennert had discovered alchemy.

The shift in Sennert's view toward alchemy had already occurred by 1611, for in that year he published his hefty *Institutiones medicinae*, which contains a long section entitled *De operationibus ad Pharmacopoeiam necessariis* (On the Operations Necessary to Pharmacy), largely devoted to chymistry. The text begins with a consideration of the etymology and history of chymistry, then moves on to a description of individual operations. Sennert gives detailed explanations of procedures such as trituration (grinding to a fine paste or powder), solution, calcination, precipitation, coction, filtration, distillation, sublimation, coagulation, and digestion. Significantly Sennert's main sources for this section appear to be Andreas Libavius and Geber: he cites the latter at least eighteen times as an authority on chymical matters in the space of fifty-six pages.[13]

---

11. Newman, *Promethean Ambitions*, chapter 2, and "Technology and Alchemical Debate, pp. 423–445. The centrality of alchemy in such discussion even in the seventeenth century is acknowledged by Sister Mary Richard Reif, "Natural Philosophy in Some Early Seventeenth Century Scholastic Textbooks" (Ph.D. diss., St. Louis University, 1962), p. 238: "One final question briefly touched upon by several authors concerns the possibility of producing a truly natural product by means of human skill. The question is usually posed in this way: 'Can art effect certain works of nature?' The specific problem which they almost always have in mind is the transmutation of baser metals through the art of alchemy."

12. Sennert, *Epitome* (1618), p. 408: "Neque hoc saltem aquae naturales praestant, sed & arte idem fieri potest."

13. Sennert, *Institutiones medicinae*, pp. 1044, 1050, 1051 (four times), 1052 (twice), 1056, 1065, 1068, 1073, 1078, 1079, 1080 (three times), and 1082.

Sennert's 1611 treatment of chymistry has a distinctly corpuscularian emphasis. Following the tradition established by Libavius's *Alchymia* of 1606, Sennert asserts that the operations of the chymical laboratory are capable of being explained in terms of *diakrisis* and *synkrisis*.[14] As we proceed through the text, Sennert gives example after example of such microscopic activity as it is effected by laboratory processes. His usual terms for the particles undergoing separation and association are the *partes minimae, partes subtiles,* and *partes grossae* of Geberian alchemy.[15] In at least two instances, however, he comes close to identifying expressly with the atomists but draws back at the crucial moment. The first of these cases is found in Sennert's discussion of calcination, where a metal is burnt to a powder:

> Geber, in Book 1, *Summa perfectionis*, Chapter 51, defines calcination as the pulverization of a thing by fire, through the privation of the humidity glueing its particles together.... But calcination has a great use in metals and the like. First, some are calcined so that they become more suited for solution. For since the structure of such bodies is harder, their essences cannot be extracted unless they be practically reduced into minimal parts and atoms.[16]

The idea here is that metals can be dissolved most easily if they have been reduced into a fine powder by calcination. The particles that result are so fine that they are almost atomic—*quasi atomos*. A similar reference to atoms occurs when Sennert discusses the alchemical process of sublimation, whereby a dry substance is vaporized and then recondensed in a sort of still called an alembic. Here Sennert again takes a corpuscularian explanation from Geber and compares it to an outright atomistic one: he says that the condensed particles collect on the side of the alembic "in the manner of atoms" (*instar atomorum*).[17] One can see, then, that Sennert in 1611 was fully aware of the advantages of corpuscular explanations, but that he was not yet willing to identify himself with the school

14. Sennert, *Institutiones*, p. 1046. Sennert refers to Libavius's *Alchymia* on pages 1043 and 1048.
15. Sennert, *Institutiones*, pp. 1046, 1050, 1081, and passim. *Partes minimae* means "very small particles," *partes subtiles* means "subtle" (i.e., small) "particles," and *partes grossae* means "gross" (i.e., large) "particles."
16. Sennert, *Institutiones*, p. 1050: "Geber, *lib. I. summ. perf. c. 51.* definit calcinationem, quod sit rei per ignem pulverisatio, per privationem humiditatis partes coniungentis.... Magnum autem habet usum calcinatio in metallis & similibus. Primo, quaedam calcinantur, ut ad solutionem fiant aptiora. Cum enim compages corporum talium fit durior: aliter essentiae inde extrahi non possunt, nisi in minimas partes & atomos quasi redigantur."
17. Sennert, *Institutiones*, p. 1079.

of Democritus. This situation changed radically by 1619, however, for in that year Sennert published his defense of chymistry, the *De chymicorum cum Aristotelicis et Galenicis consensu ac dissensu*. Here Sennert openly committed himself to the Democritean position, saying that the opinions that he has proposed are without doubt "the opinion of Democritus himself, who said that all things are composed of atoms, and that generation and corruption are nothing but *synkrisis* and *diakrisis*."[18] He would reaffirm this belief with equal vehemence in private communications as well during the early 1620s.[19]

It is important to synopsize Sennert's early development as clearly as possible, since a number of scholars have erred in thinking that his commitment to atomism came only shortly before his death.[20] In 1600, the Wittenberg academician was a self-styled though temperate acolyte of Aristotle and an outright opponent both of alchemy and atomism. By

18. Sennert, *De chymicorum* (1619): p. 358: "Atque haec, quam proposuimus, est proculdubio antiquissimorum Philosophorum de mistione opinio, & ipsius Democriti, qui ex atomis omnes componi, & generationem nihil aliud, nisi σύγκρισιν & διάκρισιν, esse statuit."

19. See Sennert's letters to his friend Michael Döring, written on 8 January and 23 March 1623, and printed in Sennert, *Danielis Sennerti Vratislaviensis... operum in sex tomos divisorum* (Lyon: Joannes Antonius Huguetan, 1676), vol. 6, pp. 589–590 and 592. The former of the two letters makes Sennert's position as a self-styled Democritean atomist absolutely clear: "Atomorum tamen & συγκρίσεως & διακρίσεως Democriticae curam plane deponere non possum; & omnino puto diligentissimo rerum naturae Scrutatori ab Aristotle factam iniuriam, & Interpretes Aristotelis cum omnia ex qualitatibus primis deducere nos iusserunt, in densissimum ignorantiae lutum, nos demersisse, a quo vix pedes extrahere possumus. Si unum saltem mensem aliis negotiis subtrahere possem, darem operam, ut Democritum redivivum exhiberem, non sine, ut spero, posteritatis commodo."

20. The incorrect claim that Sennert became an atomist only in his 1636 *Hypomnemata physica* is already found in Hans Kangro, *Joachim Jungius' Experimente und Gedanken zur Begründung der Chemie als Wissenschaft* (Wiesbaden: Franz Steiner, 1968), p. 158, n. 248: "[M]üssen wir bemerken, daß sich SENNERT erst 1636 für das Fortbestehen der Teilchen (minima corpuscula) im mistum unter ihren eigenen formae entscheidet." The same mistake is repeated by Christoph Meinel, "Early Seventeenth-Century Atomism: Theory, Epistemology, and the Insufficiency of Experiment," in *Isis* 79 (1988): 68–103; see p. 95, where Meinel says that Sennert became an atomist only "in the *Hypomnemata physica* of 1636." The error is again reproduced by Emily Michael, "Daniel Sennert on Matter and Form: At the Juncture of the Old and the New," in *Early Science and Medicine* 2 (1997): 272–299; see pp. 290–291, where she wrongly contrasts the matter theory of Sennert's 1619 *De chymicorum* with the "thoroughgoing corpuscular matter theory" of the *Hypomnemata* (I note that she seems to have modified her view in "Sennert's Sea Change," p. 31). All of these authors overlook the fact that Sennert clearly allies himself with Democritus's view "ex atomis omnes componi" on p. 358 of the *De chymicorum* of 1619. Perhaps these authors are influenced by the fact that Sennert published his manifestly nonatomist *Epitome naturalis scientiae* in 1618, only a year before the *De chymicorum*. But the prefatory note to the reader in the 1618 *Epitome* explicitly says that the text was written many years before and that it reflects a set of uncorrected views belonging among Sennert's juvenilia.

1611, he was no longer an opponent of alchemy and had even adopted a host of de facto corpuscular explanations therefrom, but he had by no means substituted the Abderite for the Stagirite. In 1619, finally, in his *De chymicorum*, Sennert had become an open adherent of Democritus, and the 1629 edition of the *De chymicorum* would carry Sennert's atomism even farther by developing his corpuscular explanations of 1611 into outright atomistic ones. We can see, then, that Sennert's growing acceptance of alchemy and his developing corpuscular theory were two projects running on closely parallel tracks. But where did Sennert's Aristotelianism fit within the framework of his newfound respect for Democritus? Had Sennert discarded his peripatetic roots in a wholesale conversion to Democritean atomism? The answer to this question can be found by considering Sennert's debt to Libavius.

As we saw before, Sennert was already employing the Libavian depiction of chymistry in terms of *diakrisis* and *synkrisis* in his *Institutiones* of 1611 but without professing atomism. Like Libavius, Sennert also accepts that Democritus veiled his true philosophy so that his opponents could not argue against its real sense.[21] Aristotle was among those who either misunderstood or misrepresented the Abderite, for the latter did not really believe in mathematically indivisible atoms, despite the Stagirite's arguments to the contrary. In fact, Sennert continues, the physically indivisible atoms of Democritus were not fundamentally opposed to Aristotle's own matter theory. At this point, Sennert's debt to Libavius reemerges, for the Wittenberg physician now claims that Democritean atoms are actually mixts:

> And when he [Democritus] says that generation comes about by the coming together (*concretione*) of these corpuscles [i.e., atoms], he does not deny mixture, but only wishes this—either that the elements do not penetrate one another, or that there does not always have to be a return to the elements and prime matter in mixture, but that new mixts can be generated from corpuscles previously mixed and already established in their own essence.[22]

21. Sennert, *De chymicorum* (1619), p. 359: "Tradiderunt nimirum Philosophi prisci sua dogmata sub verborum involucris: unde postea factum, ut adversarii eorum, contra verba potius, quam mentem illorum, quam vel non intelligebant, vel intelligere nolebant, scriberent, ipsisque, quae nunquam senserunt, affingerent."

22. Sennert, *De chymicorum* (1619), pp. 359–360: "Et cum dicit, concretione istorum corpusculorum generationem fieri, non negat mistionem, sed saltem hoc vult, vel non penetrare se Elementa, vel in mistione non semper ad elementa & materiam primam usque recurrendum esse, sed ex corpusculis jam antea mistis & in sua essentia constitutis nova mista generari posse."

Like Libavius, Sennert forsakes the genuinely Democritean idea that atoms are differentiated only by size, shape, and their spatial orientation when grouped together. He abandons the substantial uniformity of all Democritean atoms and argues that different sorts of atoms have their own essences. In another part of the 1619 *De chymicorum* he elaborates on this idea at length, arguing like Libavius that God created the atoms *ab initio* as mixtures.[23] Later he would say that these were "minima of their own genus," having distinct essences that mirrored the substantial diversity of the phenomenal world.[24] But portraying Democritus as a believer in Aristotelian mixture and essences is only the first step in Sennert's *rapprochement* between the Abderite and the Stagirite. He then proceeds to develop Libavius's idea that Aristotle himself was the proponent of a corpuscular philosophy expressed above all in the *Meteorology*:

> In fact Aristotle himself clearly cannot reject *diakrisis* and *synkrisis* in the generation of things, since he teaches *passim* that things are generated by *systasis* [coming together], *apokrisis* [separating], and *pixis* [coagulation]. And in 1 *Meteorology*, Chapter 9, he can expressly write that the particular mover in generating meteors is the motion of the sun by dissociating (διακρίνουσα) and associating (συγκρίνουσα).[25]

Sennert's discovery of the participial forms of *diakrinein* and *synkrinein*—upon which the nouns *diakrisis* and *synkrisis* are based—at *Meteorology* I 346b21-22 provides him with a linguistic middle term by which to link Aristotle to atomism. And in all fairness, Aristotle does say there that "as it approaches or recedes the sun produces dissolution (διακρίνουσα) and composition (συγκρίνουσα) and is thus the cause of generation and destruction" (Lee's translation). Hence Sennert could justly argue that *diakrisis* and *synkrisis* were fundamental processes in the physics of

23. Sennert, *De chymicorum* (1619), p. 273.

24. Sennert, *Hypomnemata physica* (1636), p. 94: "Cum vero atomorum non sit unum genus, sed pro corporum naturalium varietate varia; eas & secundum simplicia corpora, quae elementa dicuntur, & secundum composita, considerare libet. Primo enim ipsa elementa in talia corpora resolvuntur, & corpuscula rursum coeuntia, tum composita corpora, tum ipsam molem elementorum constituunt;" p. 142: "Ideoque etiam corpora illa, ex quibus haec fiunt, etsi minima dicuntur, tamen absolute talia non sunt, sed sui generis minima, id est, talia, in quae corpora illa, cum resoluuntur, abeunt, atque ita non in elementa, sed in ea, e quibus proxime constant."

25. Sennert, *De chymicorum* (1619), p. 360: "Imo ipse Aristoteles διάκρισιν & σύγκρισιν in rerum generatione plane rejicere non potest; Cum passim res συστάσει, ἀποκρίσει καὶ πίξει generari doceat. Et I. Meteor. cap. 9 expresse scribat, quod praecipuum movens in generandis meteoris sit ἡ τοῦ ἡλίου φορὰ διακρίνουσα καὶ συγκρίνουσα, Solis latio, quae segregat congregatque."

Aristotle, since at *Meteorology* I 346b21–22 they were said to underlie generation and corruption.

We are now in a position to address the question that we posed some lines ago—did Sennert's growing allegiance to Democritus entail a rejection of Aristotle? As we have just seen, Sennert followed the lead of Libavius in making Democritus into a sort of Aristotelian and Aristotle into a quasi-Democritus: the answer then is clearly negative, at least for the Sennert of 1619. But we are left with another obvious question. Did Sennert uphold the strange "atomism" of Libavius himself, with its *semina* atoms that could only acquire palpability by associating with the grosser four elements? And did Sennert believe that these same gross elements could undergo the perfect mixture described by Libavius, while the "atoms" somehow remained distinct and hidden within them?

Although we noted above that the Sennert of *De chymicorum* did believe his atoms to be mixts—and hence endowed with substantial forms—he explicitly rejected the view that *semina* lay hidden within the four elements.[26] More important, his corpuscular theory differs from that of Libavius in one respect that is absolutely fundamental. Unlike Libavius, Sennert explicitly sides with the Democritean view that "all things are made from atoms"—hence he cannot maintain that a nonatomic group of four elements exists alongside and independent of the atoms.[27] Indeed, Sennert is often explicit in arguing that the four elements precede the formation of specified substances such as the *tria prima* of mercury, sulfur, and salt. In 1619, he seems even to have believed that the elements underwent a type of Aristotelian mixture during the Creation of the world that resulted in an alteration of their forms, with the result that the three Paracelsian principles came into being.[28] He would later deny this view in the 1629 edition of *De*

---

26. Sennert, *De chymicorum* (1619), p. 225: "Verum in eo Chymicis non assentimur, quod haec semina & astra in elementis illis, quae statuunt, invisibilibus, tanquam in patria, pura & sola habitare ac beata quiete frui & hinc corporibus misceri statuunt."

27. Sennert, *De chymicorum* (1619), p. 358: "Atque haec, quam proposuimus, est proculdubio antiquissimorum Philosophorum de mistione opinio, & ipsius Democriti, qui ex atomis omnes componi, & generationem nihil aliud, nisi σύγκρισιν & διάκρισιν, esse statuit."

28. Sennert, *De chymicorum* (1619), pp. 272–274. See p. 273: "Sed cum res crearet DEUS Optimus Maximus Verbo, elementa miscuit, & mistis suas formas peculiares, ac seminales rationes & essentias Coelo analogas indidit." The idea that the elements undergo a true *mixis* in forming the principles is in accord with Sennert's position elsewhere in the *De chymicorum* of 1619. He says (p. 357) that the qualities of the elements probably act on one another to form a true mixture, but that the elemental forms are not entirely abolished—otherwise a corruption would have taken place rather than a mixture. Immediately after this, he refuses to argue against the

chymicorum, explicitly opting for the position of the eleventh-century philosopher Avicenna. The Persian philosopher had argued that mixture occurs only when the four primary qualities of the elements, hot, cold, wet, and dry, experience a "remission" (*remissio*) and become "broken" or "remitted" qualities (*qualitates fractae* or *qualitates remissae*). In Avicenna's system, the elements themselves remain unchanged, retaining their substantial forms while their broken qualities interact with one another to form a medial state, which is called a "complexion" (*complexio*). Yet the qualities of the elements do not make the new substantial form of the mixt—rather, they prepare the way for its reception, for according to Avicenna, the substantial forms of things descend from the heavens, bestowed by a supramundane "giver of forms" (*dator formarum*). It is not the hypothesis of a "giver of forms" that appealed to the mature Sennert, however, but rather Avicenna's claim that the forms of the elements themselves undergo no alteration during the process of mixture, remaining intact to form the lowest compositional stage of matter.[29]

It was not difficult for Sennert to see a form of atomism in Avicenna's stark claim that the elements persist with their substantial forms fully intact in a mixture. But it is equally important to note that Sennert's public conversion to atomism came in 1619, well before his adoption of the Avicennian view. It is a clear mistake to argue, as some scholars have done, that Sennert's confession of atomism depended on his conversion to the Avicennian perspective that the four elements remain unchanged in a mixture.[30] Although the two issues were related in Sennert's mind, it was quite possible for him to become an "atomist" without arguing that

---

Averroist position that the four elements undergo a formal remission during mixture (see pp. 357–358): "An vero formae integrae maneant, vel refringantur, quod Avenrhois & Zabarella sentiunt, aliis discutiendum relinquo. Hoc certum est, Mistum quodlibet in ea, e quibus primo constitutum est, resolvi potest: & proinde formas elementorum non aboleri. Alias enim in resolutione & putredine fieret nova elementorum generatio." In later works, such as the *Hypomnemata physica* (1636), Sennert would explicitly treat the elements themselves as atoms. It is likely that his earlier position derived from his debt to Libavius, from whom he subsequently deviated.

29. Anneliese Maier, *An der Grenze von Scholastik und Naturwissenschaft* (Rome: Edizioni di Storia e Letteratura, 1952, 2. Auflage), pp. 23–28.

30. See, for, example Michael, "Sennert on Matter and Form," pp. 286–291. In her later "Sennert's Sea Change: Atoms and Causes," (in Lüthy, Murdoch, and Newman, *Late Medieval and Early Modern Corpuscular Matter Theories* p. 21), Michael comes closer to the truth, citing a passage from the *De chymicorum* where Sennert rejects the Averroistic view of mixture in favor of the Avicennian one. But having relied on one of Sennert's collected *Opera omnia* rather than using the original first and second editions, she is apparently unaware of the fact that this passage does not occur in the 1619 edition of the *De chymicorum* but is an addition found in the 1629 version. If one compares pp. 286–287 of *De chymicorum* (1619) to pp. 152–153 of *De chymicorum* (1629), one will

the four elements themselves were atomic. While admitting the existence of the four elements, Sennert in effect wrote them out of the picture by arguing that they were not responsible for the multifarious qualities of the phenomenal world.[31] The matter was different, however, for the three chymical principles. The latter substances are "principles," both because they are property bearers responsible for a host of perceptible qualities and, more importantly, because nature allows the chymist to isolate them but prevents him from resolving them into their elementary components.[32] Like Geber, Sennert has an operational atomism that relies on the analytical tools of the laboratory to have the final say in determining the permanence of substances.

It is at this point that we may return to one of our initial premises— that the corpuscular theory obtained by fusing alchemy and the type of Aristotelianism found in the *Meteorology* was genuinely experimental. In effect, Sennert considered his principles to be the limits attained by the analytical methods of the laboratory, a concept that modern scholars have found in the work of Robert Boyle. As Bernadette Bensaude-Vincent and Isabelle Stengers have pointed out, Boyle's closely related definition of an element as that into which bodies are ultimately resolved, was based on "a "negative-empirical concept" . . . that reflected the limits of technical analysis."[33] Precisely this attitude underlay the tradition of scholastic alchemy appropriated by Sennert and developed further by

find the added passage, which calls the view of Averroes and Zabarella a "merum figmentum." In 1619, Sennert had still not made up his mind about the permanence of the elemental forms despite the fact that he had become a self-professed atomist.

31. Sennert, *De chymicorum* (1619), p. 283.

32. Sennert, *De chymicorum* (1619), p. 282: "Neque enim quod saepe jam monuimus, cum Paracelsistis facimus, qui statuunt ista tria principia esse planè prima, vel etiam ante elementa: sufficiat, si obtineamus, esse post elementa, et talia à quibus pleraeque facultates, quas medicus in medicamentis expetit, oriantur, & ultra quae resolvere Corpora naturalia non tentat Chymicus: Cum nec natura hic facilem sese in resolvendo, si quis ulterius progredi cupiat, praebet: Nisi forsan destruere plane potius, quam resolvere res cupiat." The destruction referred to here is not a resolution into the elements, but rather a hypothetical contrary-to-fact destruction of matter. That Sennert really views the resolution into atomic *prima mixta* as often being the ultimate obtainable in the laboratory is corroborated on p. 285: "Ultimam tamen, cum dicimus resolutionem, non eam plane intelligimus, quae fit in elementa: Cum natura tam facilem in resolvendo sese non semper praebeat: Et in hisce principiis Chymico saepe subsistendum sit" [But when we say "the final resolution," to be sure we do not understand a resolution into the elements. For Nature does not always show herself so easy in resolving, and for the chymist there must often be a limit in these principles].

33. Bensaude-Vincent and Stengers, *History of Chemistry*, p. 37. See Thackray, *Atoms and Powers*, p. 168, who in turn cites David Knight for this concept. For Boyle's definition of "element," see *The Sceptical Chymist*, in Hunter and Davis, *Works*, vol. 2, p. 345. Boyle himself of course

his heirs. This approach to material composition would have great consequences in the history of later chemistry. It resurfaces, for example, in Lavoisier's claim a century after Boyle that the term "element" should be restricted to "the last point which analysis is capable of reaching ... all the substances into which we are capable, by any means, to reduce bodies by decomposition."[34] The same combination of dogged empiricism and faith in the analytical capabilities of chemistry underlies the "chemical atomism" of the nineteenth century, with its resolute refusal to speculate about the microstructure of the undivided elements that made up its stock and trade.[35] Yet this was hardly an idea that originated with Dalton, Lavoisier, or Boyle. As Sennert himself points out, the validity of chymical analysis is borne out by the scholastic "axiom" *in quae dissolvi possunt composita, ex iisdem coaluerunt* ("The things into which composites can be dissolved are the things out of which they are made"—based on Aristotle's *De caelo* 3 302a 15–18). In the context of atomism, the analytical agents of the chymist are used by Sennert to define the constitution of "indivisibility."[36]

As we have seen, Sennert's operational approach had already been taken by Geber and his multitude of followers but generally without an

---

had serious doubts about the ability of commonly available tools of analysis (as opposed to the alkahest) to reveal the genuine constituents of bodies. See William R. Newman and Lawrence M. Principe, *Alchemy Tried in the Fire* (Chicago: University of Chicago Press, 2002), pp. 273–314.

34. Antoine Laurent Lavoisier, *Elements of Chemistry*, trans. R. Kerr (Edinburgh, 1790), p. xxiv.

35. Here it is apposite to quote David Knight's description of chemical atomism: "[A]ll that it was necessary to know was that some substances could not be further analysed. These were the elements; and the smallest units of them appearing in chemical reactions were the atoms. With the progress of science, the list of elements might change as some yielded to more powerful techniques. In using the term 'atom' one was not making any statements about indivisibility." From Knight, *Atoms and Elements: A Study of Theories of Matter in England in the Nineteenth Century* (London: Hutchinson, 1967), p. 12. This statement could be applied without anachronism to the corpuscular *partes* of Geber or the *prima mista* Sennert, both products of chymical analysis. Alan Rocke's important book *Chemical Atomism in the Nineteenth Century: From Dalton to Cannizaro* (Columbus: Ohio State University Press, 1984) emphasizes the importance of atomic weight for the development of nineteenth-century chemical atomism. Obviously this factor distinguishes the chemistry of John Dalton and his heirs from the chymistry of the seventeenth century, but the operational consideration of an atom as that which resists laboratory analysis remains the same. See Rocke, *Chemical Atomism*, pp. 1–15.

36. Kangro interestingly notes a similar approach on the part of Sennert's intellectual heir, the Hamburg academician Joachim Jungius. Kangro refers to Jungius's "Prinzip der Aussagebeschränkung" and links this conservative approach to Jungius's attempt to determine the material elements of things by a posteriori rather than a priori means. The German historian does not make a point of connecting Jungius's method to Aristotelianism or to Sennert, however. See Kangro, *Joachim Jungius' Experimente*, pp. 206–212.

explicit adherence to the notion of the indivisible connoted by the term "atom."[37] Geberian alchemy postulated at least three states in which corpuscles could exist. First, there were the individual elementary *partes*, hypothetical bits of fire, air, water and earth that played almost no explanatory role other than that of providing a sort of substrate. Following the general outline provided by Aristotelian matter theory, Geber believed that the elementary corpuscles then went into combination to form the larger *partes* of sulfur and mercury capable of being analyzed out of metals by means of laboratory procedures.[38] In accordance with the scholastic principle that "the things into which composites can be dissolved are the things out of which they are made," these aggregate corpuscles in turn combined to form the minimal parts of the metals themselves. As we showed in our treatment of the *Summa perfectionis*, Geber used laboratory analysis in addition to confirm that his aggregate corpuscles were homoeomerous (in the restricted, corpuscular sense in which he used that term), and further experiments allowed him to argue that the metals were composed of both volatile and fixed sulfur and mercury in varying proportions. What Geber did not do, beyond making an explicit appeal to atomism, was to employ the much more powerful tools of analysis at Sennert's disposal. Nor could he have done so, for between the time when the *Summa perfectionis* was written and the seventeenth century, the mineral acids—sulfuric, hydrochloric, nitric, and the mixture of the latter two, called *aqua regia*, had been discovered. These remarkable chemicals were able to dissolve and separate the metals with a speed and activity undreamt of by the author of the *Summa perfectionis*.[39] They would provide Sennert with his most impressive tools for enacting the reduction to the pristine state.

37. Although the major trend in Geberian alchemy was corpuscularist rather than atomistic, some followers of Geber did in fact commit themselves to an openly atomist perspective. See Newman, "Experimental Corpuscular Theory in Aristotelian Alchemy: From Geber to Sennert," in Lüthy, Murdoch, and Newman, *Late Medieval and Early Modern Corpuscular Matter Theories*, pp. 301–306.

38. I refer, principally, to the process of calcination, which Geber interprets as the exhalation of a given metal's sulfur. What is left at the end of the process, according to him, is a powdery, fixed mercury. From a more modern perspective, this explanation is illusory, of course, since the process involved is actually one of oxidation. What is significant for us, however, is Geber's claim that such a process of laboratory analysis reveals the components of the metals.

39. Robert Multhauf, "The Relationship Between Technology and Natural Philosophy, ca. 1250–1650: as Illustrated by the Technology of the Mineral Acids" (Ph.D. diss., University of California, 1953).

JULIUS CAESAR SCALIGER'S REFORMULATION OF MIXTURE THEORY

Already in the 1619 *De chymicorum*, Sennert presents the following example of his famous proof for the existence of atoms by means of a reduction to the pristine state (*reductio in pristinum statum*). The demonstration relies on nitric acid as a means of separating silver from an alloy of that metal and gold (often called electrum):

> If gold and silver melt together, they are so thoroughly mixed *per minima* that the gold cannot in any way be detected by sight, but if aqua fortis is then poured on, the silver is so thoroughly dissolved that no metal can be detected in the water by sight. But since it is really present, it can emerge thence in segregated form, and certainly in such a way that both the gold and the silver retain their own nature; and it is in this fashion collected into the subtlest calx, which is nothing other than a heap of innumerable atoms, which is again reduced into the purest gold and silver by fusion.[40]

Although Sennert is implicitly arguing here against a complex tradition of explaining mixture in terms of the generation and corruption of forms, the empirical basis of his atomistic assertion is easily grasped. First, and most important, the silver has been so thoroughly combined with the gold and then dissolved by nitric acid that it is no longer perceptible. And yet, despite having been subjected to one of the most powerful agents of analysis available in the seventeenth century, the metal can be regained intact by means of precipitation. From the perspective of the "negative-empirical" principle, it is therefore operationally *a-tomos*—indivisible—since it has resisted all efforts at laboratory decomposition into its components. Second, the precipitated silver particles are so small that they satisfy another canonic criterion of atomism—the requirement of minute size.[41] In his later works, Sennert would underscore the minuteness of

40. Sennert, *De chymicorum* (1619), p. 362: "Si aurum & argentum simul liquescant, ita per minima miscentur, ut visu deprehendi aurum in argento nullo modo possit: si vero postea aqua fortis affundatur, ita solvitur argentum, ut ullum metallum in ea aqua deprehendi visu non possit: cum tamen revera insit & hinc segregatum emergat; & quidem ita, ut & aurum & argentum suam naturam retineat; & hoc modo in subtilissimam calcem, quae nihil aliud est, quam congeries aliqua innumerabilium atomorum, redigatur, quae in aurum & argentum purissimum fusione iterum reducitur."

41. The precipitate would actually not be pure silver but silver carbonate, which would reduce into metallic silver upon heating and fusion. This is immaterial to Sennert's point, however, which is that the silver itself has not been broken down either by its mixture with the gold or with the acid.

the silver atoms by passing them through filter paper before their precipitation from the acid-silver solution.[42]

Yet Sennert was by no means the first to use the reduction to the pristine state in order to demonstrate that the generation and corruption of substantial forms was unable to explain the dissolution and recapturing of metals. As we have seen in chapter 1, a version of this argument already appeared in the medieval *Theorica et practica* by Paul of Taranto. Although the German academic does not seem to have known Paul's *Theorica et practica*, Sennert was also preceded in this demonstration by the famous Aristotelian scholar Julius Caesar Scaliger, whose *Exotericarum exercitationum* (1557) was a favorite work of Sennert's since his days as an undergraduate at Wittenberg. Despite the fact that Scaliger's hugely unsystematic work was written as a critique of Girolamo Cardano's *De subtilitate*, it contained a theory of mixture that would exercise a profound influence on later corpuscular theorists.[43] Sennert himself had used the *Exotericarum* extensively in composing his 1599–1600 *Epitome naturalis scientiae*, but at this prealchemical stage of his career, the young Sennert was still wedded to the more orthodox mixture theory of the twelfth-century philosopher Averroes, which asserted that the ingredients of a genuine mixture had to undergo the "union" described in *De generatione et corruptione*, to become a totally homogeneous body.[44] It was not until Sennert's appreciation of chymistry had developed that he fully grasped the importance of Scaliger's ideas on this subject and rejected the theory of Averroes.

---

42. Sennert, *De chymicorum* (1629; 2nd ed.), p. 393: "Cum enim spiritus salsi & sales, dum corpora metallorum solvunt, iis per minima permisceantur, & admistione spirituum salsorum ac salium corpora haec in minimas atomos redigantur, ita quidem ut metalla in aquis fortibus & regiis soluta per chartam transeant: sale illo alieno separato suae pristinae formae restituuntur; & licet in forma pulvis relinquantur, igne fusorio tamen pristinam formam facile acquirunt." Sennert, *Hypomnemata physica* (Frankfurt: Schleichius, 1636), pp. 109–110.

43. Christoph Lüthy, "An Aristotelian Watchdog as Avant-garde Physicist: Julius Caesar Scaliger," in *Monist* 84 (2001): 542–561.

44. Sennert, *Epitome* (1599), *Disputatio* XIV, *Thesis* XIX: "Sed non levis hic oritur controversia, quomodo elementa maneant in mistis. Nam cum miscibilia debeant esse alterata, non corrupta; videtur, quod elementa in misto remaneant, non pereunt. Contra cum mistio non sit mera congregatio tantum miscibilium, sed unio: ita ut ex pluribus naturis una fiat, videntur elementa interire, & inde una quaedam nova natura fieri: de qua re variae variorum sunt sententiae. Nos, caeteris rejectis, Avenrois sequemur, qui putavit, non solum Qualitates, sed ipsas etiam formas Elementorum manere in misto, refractas tamen, ita, ut ex omnibus una forma fiat, non tanquam a termino a quo, sed tanquam ex partibus, non aliter, ut ex coloribus & saporibus extremis fiunt medii; ita, ut qui fuerunt gradus formarum Elementorum, jam fiant gradus formae misti." For the theory of Averroes, see Maier, *An der Grenze*, pp. 28–31.

In *Exercitatio* 101, Scaliger presents his own definition of mixture, which would become increasingly celebrated in the following decades— "Mixture is the motion of minimal bodies up to [their] mutual contact, so that there is union."[45] With these words, Scaliger added the important concepts of motion, minimal bodies, and contact to the Aristotelian definition of mixture found in *De generatione et corruptione* (Mixture is the union of the altered miscibles).[46] Although Scaliger's definition may at first sound atomistic, he was eager to dissociate himself from the atomism of antiquity. Immediately after defining mixture, Scaliger said "Nor do our corpuscles touch one another like Epicurean atoms, but rather as a continuous body, so that one thing results. For it becomes one by means of a continuity of boundaries, which is common to all mixts."[47] Although these words admit a host of problems, Scaliger's following examples suggest strongly that he had in mind a type of corpuscle that was free-form and flexible. His major reason for rejecting the atoms of Greek antiquity seems to be that they would not allow for genuine "continuity of boundaries," being rigid, oddly shaped, and separated by interstitial gaps. His own corpuscles, although they do not interpenetrate, can at least conform to one another's boundaries so that no other corpuscles can get between them. In a loose sense Scaliger's corpuscles could therefore be said to satisfy the Aristotelian criterion of "union" insofar as there was nothing separating them from one another. Needless to say, this solution raises serious philosophical difficulties—why, for example, would we call Scaliger's mixture a state of continuity rather than mere contiguity? Nonetheless, it does seem to have been his view, as his following elaborations reveal.

Scaliger goes on to analyze mixture into its various types. He excludes two kinds of "mixture" from serious consideration. The first is a purely "metaphorical" mixture, such as that which occurred in the family tree of the Scalas between Germans and Italians. The second is the type of mixture that Aristotle called *synthesis*, as when millet and beans are shaken in a jar. This is not true mixture, Scaliger says, for the ingredients are

45. Julius Caesar Scaliger, *Exotericarum exercitationum liber xv* (Lyon: Vidua Antonii de Harsy, 1615), p. 273: "Mistio est motus corporum minimorum ad mutuum contactum, ut fiat unio."

46. Aristotle, *De generatione et corruptione*, in *Aristotelis opera cum Averrois commentariis* (Venice: apud Junctas, 1562; Minerva reprint, 1962), vol. 5, book 2, chapter 10, fol. 370v: "Mistio autem est miscibilium alteratorum unio."

47. Scaliger, *Exotericarum exercitationum*, p. 273: "Neque enim velut atomi Epicureae sese contingunt: ita corpuscula nostra, sed ut continuum corpus, atque unum fiat. Fit enim unum continuatione terminorum: quae est mistis omnibus communis."

dry and do not cohere, "for each grain is and remains a certain whole unto itself, and is not joined (*continuatur*) with another to make one body."[48] Here it is important to note Scaliger's emphasis on the dryness of the ingredients in such spurious mixture: it is this that prohibits their cohesion and concomitant "union."

After saying what mixture is not, Scaliger then proceeds to describe three types of genuine mixture that conform with his own definition. The first sort of genuine mixture is the type where the mixed ingredients can again be separated. Unlike Erastus and the overt followers of Thomas Aquinas, Scaliger has no problem with the idea that the ingredients of a genuine mixture can again be regained intact. Citing Pliny, he says that a mixture of wine and water can be separated by placing it in a container made of ivy-wood.[49] The second type of mixture occurs in the case of the so-called "imperfect mixts." In traditional fashion, Scaliger divides these into two types—purely meteorological products, such as snow, hail, and so forth, and the more durable mixtures found in some clods and stones. The third type of mixture, finally, is that found in animate bodies.

Scaliger then asks: are all three types of mixture really the same, and if not, how do they differ? His answer is the following—in the first two types of mixture, the corpuscles remain distinct and have their own forms, whereas the third involves an actual formal unification. Hence water is mixed with wine in a very different way from the sort of mixture that occurs when the elements combine in the body of a lion. In cases of inanimate things where the individual corpuscles seem to blend together, they are really forming a *crama* (from the Greek κρᾶμα), which is "not one by form, but only by interconnection (*continuatione*)." In the case of a *crama*, there are as many particles as there were in the individual ingredients before their mixture. But where are these parts? Do they continue to exist in act? Here we begin to encounter conceptual difficulties. Scaliger responds as follows—"They are not present in mathematical act, for they are not delimited and separated by borders, but they are present in natural act." The point of this is that the individual parts of the *crama* retain their own forms; hence they are present in "natural act." The difficulty occurs with Scaliger's claim that they are not present in "mathematical act." If this is literally the case, how can we think of spatially distinct

48. Scaliger, *Exotericarum exercitationum*, p. 273: "Unumquodque enim granum sibiipsi totum quoddam & est, & manet: neque cum alio continuatur ad unum corpus efficiendum."
49. Pliny, *Natural History* (Cambridge, MA: Harvard University Press, 1968), book 16, chapter 63, p. 488 (Latin), 489 (English). The same claim is made by Cato, *On Agriculture* (Cambridge, MA: Harvard University Press, 1935), chapter 111, pp. 100–103.

parts of matter acting as vectors for their own forms? The safest way to construe Scaliger's meaning is by assuming that he means his corpuscles to lack permanently defined boundaries. They are amoebalike bits that conform to one another's shapes in order to produce the appearance of uniformity.

Since a *crama* is not per se one but rather a collection of particles with their own forms united *per accidens*, it should be possible, Scaliger says, to separate the ingredients from one another. At this point, then, he introduces the alloy of gold and silver that Sennert would later employ to much wider effect.

> Therefore the parts preserved under their own form can be separated from the whole, and can proceed separately to a union [with one another] by virtue of their own form. This appears quite clearly in gold and silver, separated by the faculty of that marvelous water [that is, "parting waters"—mineral acids]. Although they appear to our eyes as if under one form, this is effected by the subtlety of art, so that it is necessary to understand that there was not one natural body but two, as they are afterwards revealed. Which would certainly be impossible unless the forms themselves were integral, [and] the parts themselves were preserved in act under those forms. If the parts were dry, they would create no difficulty in our minds, but because they are liquid, they give up one thing, namely, the limiting of boundaries, while they retain another, namely, the forms themselves.[50]

To summarize Scaliger's position, his words reveal that he viewed the individual corpuscles making up the gold and silver in his alloy as being "liquid." These amoebalike particles were bearers of the formal properties that distinguished gold and silver from one another. Although Scaliger is rather vague about details of the laboratory, it appears that he is describing the same dissolution of silver out of a gold-silver alloy by means of nitric acid that Sennert would describe in the 1619 *De chymicorum*. We may assume that Scaliger also knew, like Sennert, that a base can cause the dissolved silver to precipitate—otherwise he would have had no basis for his claim that the particles of silver were preserved in

---

50. Scaliger, *Exotericarum exercitationum*, p. 275: "Iccirco sub sua quaeque forma servatae partes de integro possunt segregari: ac seorsum sub suae formae vi coire ad unionem. Id quod clarius quoque patet in auro & argento per aquae illius mirabilis facultatem separatis. Quae cum quasi sub una forma nostris oculis apparerent, artis subtilitate effectum est, ut non unum corpus naturale, sed duo ita fuisse, uti postea visa sunt, intelligere necesse sit. Quod sane fieri non posset, nisi formae ipsae integrae, ipsae partes sub suis formis actu naturali servarentur. Quae partes, si essent siccae, nullam crearent in animis nostris difficultatem: sed quia liquidae sunt, alterum admittunt, id est extremorum praescriptionem: alterum servant, scilicet formas ipsas."

act throughout the process. But beyond explaining the fact that the alloyed gold and silver can give the illusion of being one substance and yet
can again be separated, Scaliger's corpuscles do little work for him. This
would not be the case for Daniel Sennert, who combined Scaliger's concern for the formal integrity of ingredients with the corpuscular reductionism of Geber and the *synkrisis* and *diakrisis* of Democritean atomism.

There were others, however, who integrated Scaliger's theory of
mixture into the context of alchemy before Sennert. One of these
was Nicholas Guibert, who wrote a stinging attack on chrysopoeia in
1603, largely motivated by the antialchemical work of Thomas Erastus.
Guibert's *Alchymia ratione et experientia impugnata* (Alchemy Attacked by
Reason and Experience) reveals very clearly how an Erastian outlook
could encourage a corpuscularian viewpoint, for in Guibert's work, the
approach of Erastus is combined with the rudimentary corpuscularism
of Scaliger. Much of Guibert's *Alchymia* consists of condensed Erastian
arguments. He accepts the *reditus* principle, for example—that there is
no return from privation to a habit, and like Erastus, Guibert uses this
to debunk alchemical transmutation. As he puts it, if the form of lead
is deleted from that metal, it cannot then be recombined with the remaining matter to regenerate lead. This would be just as impossible to
nature and art as the resurrection of a dead man. How then can the alchemists expect to convert lead to an even more perfect metal, such as
gold?[51] Nonetheless, Guibert was himself a medical chymist, and despite
his rejection of chrysopoeia, he had spent some forty years studying the
chymical art.[52] He felt compelled, therefore, to account for the common
chymical phenomenon of reduction to the pristine state. How could he
do this while employing the destructive hylomorphic tools of Erastus?
After all, if one interpreted the dissolution of a metal in acid as the corruption of its form, the reduction of that metal by means of precipitation
would seem directly to violate the *reditus* principle, the favored antialchemical weapon of Erastus. There was a simple escape to this problem,
however—namely, by amplifying Erastus's own claim that the substances
reduced by chymists were compositions rather than mixtures. And in this,
Guibert found himself immensely aided by Scaliger, whose discussion
from *Exercitatio* 101 he quotes almost verbatim. Like Scaliger, Guibert
accepts that the components of gold and silver are "liquid" particles that
retain their formal integrity even when they lose their visual identity in

51. N. Guibert, *Alchymia ratione et experientia impugnata* (Argentorati, 1603), pp. 8–11.
52. Guibert, *Alchymia*, p. 2.

an alloy.[53] Hence the reduction to the pristine state involves no loss and recapture of a substantial form. This answer simultaneously saves the reality of reduction while also allowing for the Erastian argument that *if* a metallic form should really be corrupted, it would be impossible to replace it with the same or another metallic form. Thus Guibert managed both to cast doubt on chrysopoeia and to save the phenomenon of reduction to the pristine state.

Guibert's arguments against chrysopoeia would in turn be debunked by Libavius in his own *Defensio alchymiae transmutatoriae* of 1604. Thanks to his love of ad hominem arguments, it is excruciatingly difficult to extract a coherent matter theory from this part of Libavius's oeuvre. Nonetheless, Libavius does make several points that are highly relevant to our discussion of the reduction to the pristine state. In essence, Libavius rejects outright the universality of the *reditus* rule that Erastus and his follower Guibert had used to invalidate the transmutation of metals. First, Libavius points out that Aristotle himself accepted that water could assume the form of air and then revert directly to water. But this return to a habit can be witnessed in more particular things as well, not just in the elements. Here Libavius brings in the evidence of the chymical laboratory—Guibert's arguments are rash, and they can be refuted by the mere reduction of quicksilver that has previously been converted to "ash," not to mention the fact that metals can be reduced after having been converted to stones, calces, and liquors.[54] Even in the case of living beings the *reditus* principle does not always hold, for wasps drowned in oil can be resuscitated in vinegar, and flies drowned in water can be brought back to life by the warmth of hot ashes.[55] Whatever the merit of these arguments, they illustrate with great clarity the opposition that existed between the *reditus* principle and its antithesis, the reduction to the pristine state.

One could certainly point to other early modern chymical writers who employed the reduction to the pristine state before Sennert, such as the Nivernais lawyer and chymist Gaston Duclo, or the writer on medical chymistry Angelus Sala. And indeed, these authors also fused earlier

---

53. Guibert, *Alchymia*, pp. 16–18.

54. Libavius, *Defensio et declaratio perspicua alchymiae transmutatoriae* (Ursel: Petrus Kopffius, 1604), pp. 28–30: "Sed temerariae sunt istae contentiones refutanturque vel solius hydrargyri in favillam versi reductione, non in speciem aliam; sed eandem quae fuerat prius. Ita Metalla re solvi possunt in lapides, calces, liquores, indeque in eadem reverti, quamquam non pari semper quantitate."

55. Libavius, *Defensio*, pp. 32–33.

Geberian corpuscular ideas with their more modern observations.[56] Yet unlike Sennert, none of these authors used the reduction to the pristine state as an explicit demonstration for the existence of Democritean atoms or argued that "all things are composed of atoms, and that generation and corruption are nothing but *synkrisis* and *diakrisis.*" It was up to Sennert to combine these different facets into a coherent experimental corpuscularism explicitly designed to replace the scholastic theories of mixture that this sheltered product of Wittenberg had himself imbibed from his earliest youth. It is time, then, to consider in some detail the theories that Sennert hoped to debunk and to determine how the reduction to the pristine state was supposed to explode these explanations of mixture forming the young Sennert's *Gedankengut.*

SENNERT'S ATOMISM AND ITS TARGETS

The primary target of Sennert's experimentation was a scholastic theory that he explicitly rejected in his *De chymicorum* of 1619, namely, that for mixture or generation to take place, "there must always be a resolution [of the ingredients] into the [four] elements; [or] indeed, as some say, all the way into the *materia prima*, and that no new mixt can be produced unless the elements are mixed de novo."[57] It is this theory that we must address if we wish to understand the goals and success of Sennert's work rather than appraising him in terms of anachronistic criteria such as the presence or absence of a quantitative method in his experimental research.[58] Yet beneath Sennert's seemingly simple words lurks a mare's

---

56. Lawrence Principe, "Diversity in Alchemy: The Case of Gaston 'Claveus' DuClo, a Scholastic Mercurialist Chrysopoeian," in *Reading the Book of Nature: The Other Side of the Scientific Revolution*, ed. Allen G. Debus and Michael T. Walton (Kirksville: Sixteenth Century Journal, 1998), pp. 181–198; Urs Leo Gantenbein, *Der Chemiater Angelus Sala* (Zurich: Juris Druck & Verlag Dietikon, 1992); and Meinel, "Early Seventeenth-Century Atomism," pp. 68–103.

57. Sennert, *De chymicorum* (1619), p. 351: "[Q]uod semper resolutionem usque ad elementa, imo etiam quidam ad materiam primam usque fieri statuunt, & nullum, nisi de novo elementa misceantur, novum mistum produci posse dicunt."

58. I have in mind the otherwise exemplary article of Meinel, "Early Seventeenth-Century Atomism"; see especially p. 78. Meinel sees Sennert's arguments as being primarily "figurative and rhetorical" and lacking "the vaguest idea of a quantitative methodology." Sennert's experiments should not be judged against the touchstone of modern quantitative methods, however, but against the theories that they set out to disprove. Meinel's inspiration, Hans Kangro's article, "Erklärungswert und Schwierigkeiten der Atomhypothese und ihrer Anwendung auf chemische Probleme in der ersten Hälfte des 17. Jahrhunderts," *Technikgeschichte* 35 (1968): 14–36, purports to do just that but fails to address the most compelling early modern problems, and instead imports extraneous ones such as the issue of *creatio ex nihilo*.

nest of commentary and debate extending back for several millennia. The basis of the argument lies in Aristotle's distinction between the process of generation and corruption on the one hand, and mixture on the other. In *De generatione et corruptione* (I 4 319b6–21), Aristotle argues that generation occurs when not just the qualities of a thing, but the sensible substrate in which those very qualities inhere, is changed beyond recognition. Hence a sick body becoming well would not involve a generation since the body remains a body whether sick or well, but air becoming water would require a corruption of the air followed by a generation of the water. So far so good, but now let us consider Aristotle's treatment of mixture. Real *mixis* for Aristotle is a state of perfect homogeneity: the initial ingredients are altered beyond recognition so that even one endowed with the superhuman vision of Lynceus would not be able to find them in the mixed substance. Does this not mean that the original ingredients, in particular the four elements out of which they were made, have been corrupted in order to form the new mixture? Could such a resolution or corruption be said to result in a destruction of the elements themselves? As we saw in chapter 1, Thomas Aquinas and his school gave an affirmative answer to this question. And yet this raised an obvious problem: if the ingredients of a mixture were totally destroyed in the process of mixing, then how could one reasonably speak of "mixture" at all, as distinct from generation and corruption per se?[59]

In the thirteenth century Thomas had adopted the position that the forms of the elements are destroyed during the process of mixture and that they remain only "virtually" (*in virtute* or *virtualiter*). A generation after Thomas, John Duns Scotus had added further refinements to this explanation of the elements' persistence in a mixture but only by explicitly severing the relation of numerical identity between the elements of the ingredients and those of the mixt. Scotus argued that for the elementary forms and qualities of the ingredients to remain *virtualiter* in the mixt meant only that there was a "similarity- and affinity-relationship between the one and the other." Any elements that might be retrievable from the mixt would not be numerically identical to those that entered it as ingredients, but would only be similar. As the historian of medieval philosophy Anneliese Maier puts it, Scotus's system led to an elimination of the causal connection between the elements and the mixture—"What remains is only an *ordo successionis* (order of succession) and a similarity among the successively following forms." The fact that philosophers of

---

59. Aristotle alludes to this problem at *De generatione et corruptione* I 10 327a35–327b6.

such eminence as Thomas and Scotus argued for a merely virtual persistence of the elements in a mixture gives a hint of the challenge that Sennert's atomism would offer to their heirs among the early modern scholastics. The issue of *mixis*, which had produced a scholastic debate of monumental dimensions stretching from the ancient commentators of Aristotle through the Arabs and their Latin successors, was still a major sticking point even in the early seventeenth century.[60]

The Thomistic theory of mixture dovetailed nicely with the belief in the "unity of the substantial form" espoused by the Angelic doctor and his followers, namely, the idea discussed in chapter 1 that every entity can have only one substantial form. The substantial form that imposes humanity on an individual human, for example, makes him something other than a hot, wet, steaming pile of earth. The persistence of the forms of the elements in the mixt would, according to Thomas, imply precisely the disunity of a heap. Sennert was aware of the fact, however, that there were other contemporary peripatetics who drew on a longstanding tradition maintaining the opposite position, a view that we already encountered in our earlier discussion of Paul of Taranto's *Theorica et practica*. According to this view, there could be a plurality of substantial forms in one substance, although the number of forms allowed varied from author to author and was often quite restricted. Already in his proto-*Epitome naturalis scientiae* of 1600, Sennert presented as correct a version of the plurality of forms theory derived from the Paduan philosopher Jacobus

---

60. A beautiful treatment of this debate in the High and Late Middle Ages can be found in Maier, *An Der Grenze*. Maier's description of the Scotist theory of mixture is worth quoting at length (pp. 107–108): "[D]as *remanere virtualiter* der Elementarformen und -qualitäten in der forma mixti und ihrer qualitas media ist nun als ein blosses Ähnlichkeits- und Affinitätsverhältnis zwischen diesen und jenen festgelegt; und die Frage, wie man unten diesen Umständen doch noch annehmen kann, dass das mixtum aus den Elementen besteht, oder dass wenigstens die forma mixti die Materie *mediantibus formis elementorum* informiert, ist beantwortet. Diese Auffassung hat eine starke Wirkung gehabt und ist für das 14. Jahrhundert zu der massgebenden geworden. Aber es ist klar, dass damit der Lehre vom Aufbau der physischen Welt aus den Elementen eigentlich schon der Todesstoss versezt ist. Um das Bild zu vervollständigen, sei noch erwähnt, dass Duns, wie die Mehrzahl der scholastiken Denker, die Entstehung der forma mixti (ihre Einführung in die Materie) auf das Wirken überirdischer Kräfte zurückführt. Es fällt also auch der Kausalzusammenhang weg, sogar die Disponierung der Materie für die Aufnahme der neuen Form durch das Wirken der Elementarkräfte. Was bleibt, ist lediglich ein ordo successionis und eine Ähnlichkeit unter den aufeinander folgenden Formen." On Scotus, see also Xaver Pfeifer, *Die Controverse über das Beharren der Elemente in den Verbindungen von Aristoteles bis zur Gegenwart, Programm zum Schlusse des Studienjahrs 1878/79* (Dillingen: Adalbert Kold, 1879), pp. 28, 37. For scholastic theories of mixture, see also Norma E. Emerton, *The Scientific Reinterpretation of Form* (Ithaca, NY: Cornell University Press, 1984), pp. 76–105.

Zabarella. Sennert's 1600 view, an abbreviation of the treatment given by Zabarella in his 1590 *De rebus naturalibus*, is as follows:

> Does any form that existed before in the corrupted thing, beyond the prime matter, remain in the generated thing? Or rather, does a resolution up to the prime matter happen in every generation? It is not difficult to see that there has to be a resolution up to the prime matter in the [substantial] mutation of the elements and in that of inanimate things, and that no form that was in the corrupted thing remains in the generated thing. For although the forms of the elements do not wholly perish in mixts, but some degrees of them remain, still they are not distinct from the form of the mixt, but gathered together they make up the form of the mixt, so that nothing may be found beyond the form of the mixt and the prime matter in the mixed body . . . But in the [substantial] mutation of animate beings a resolution up to the prime matter does not occur; rather after the death of the living being, the form of the mixture remains, and in the generation [of an animate being], the form of the mixture is produced first, rather than the soul.[61]

The essence of Sennert's position—and Zabarella's—was that animate beings, because they have a soul, must also have at least one other form beneath that entity, and this subordinate form must be a *forma mixti*. The reasoning behind this was that upon death the soul departs but the body remains intact, at least for a while. Hence one was confronted by two possibilities: either the body had had its own form all along (a *forma mixti*) distinct from the soul, or else a new form—a *forma cadaveris*—was generated instantly upon death to account for the perpetuation of the body. Zabarella and Sennert (in 1600) in fact maintain both views simultaneously, perhaps in order to reconcile their position with that

---

61. Sennert, *Epitome* (1600), *Disputatio XV, Thesis VII*: "Ex his facile responsio peti potest, ad quaestionem illam: An praeter materiam primam aliqua forma, quae prius fuerat in corrupto, remaneat in genito; vel, An in omni generatione fit resolutio ad primam materiam. Non etenim difficile est perspicere, in mutatione Elementorum, & mistorum inanimatorum fieri resolutionem ad primam usque materiam, nec formam, quae erat in corrupto, manere in genito. Nam quamvis in mistis formae Elementorum non penitus pereunt, sed eorum gradus aliqui manent reliqui: illi tamen non sunt distincti a forma misti, sed collecti unam formam misti constituunt, ita, ut nihil praeter formam misti, & materiam primam in corpore misto reperire sit: & uno misto intereunte, gradus formarum elementarium, qui manent in genito, non eandem formalitatem, ut loquuntur, seu idem Esse specificum retinent, nec distincti a forma producta servantur, sed in novae formae constitutionem conspirant, ejusque gradus, & quasi partes quaedam fiunt. In animatorum autem mutatione non fit resolutio ad materiam primam, sed & post viventis interitum remanet forma mistionis, & in generatione prius producitur forma mistionis, quam anima." This is based closely on Zabarella, *De communi rerum generatione et interitu*, in Jacobus Zabarella, *De rebus naturalibus* (Frankfurt: Zetzner, 1606; Minerva reprint, 1966), columns 394–426; see especially chapters 1–4, columns 395–408.

of earlier scholastics. They admit that a *forma cadaveris* is indeed found after death but that this is nothing more than a congeries of various other forms that already preexisted beneath the vesture of the soul. Hence the *forma cadaveris* is not really a new entity that comes into being when the soul departs.[62]

According to the Zabarellan position to which Sennert subscribed in 1600, the forms of the elements do not wholly perish in mixts, but some "degrees" of them remain. Zabarella's position was an elaboration on that of the twelfth-century philosopher Averroes, whom Sennert dutifully cited in the 1600 *Epitome* as its ultimate source.[63] Averroes had maintained that mixture occurred only when the substantial forms of the elements underwent a "remission" (*remissio*), thereby becoming "broken" (*fractae*).[64] To the objection that substantial forms cannot become more or less intense, Averroes responded that the substantial forms of the elements are not complete forms but instead something between substance and accident. This influential theory allowed Averroes to account for some permanence of the elements in the mixt while also fulfilling Aristotle's dictum in *De generatione et corruptione* that the ingredients of a genuine mixture had to undergo the "union" to become a totally homogeneous body.[65] But Averroes sparked a further debate among the Latins as to whether the new substantial form of the mixt was itself a separate entity imposed upon the preexisting broken forms of the elements or merely the sum total of the broken elements themselves. The position of Zabarella and that of Sennert in 1600 was clearly the latter. As Sennert explicitly says, the forms of the elements are not distinct from the form of the mixt, but "gathered together they make up the form of the mixt, so that nothing may be found beyond the form of the mixt and the prime matter in the mixed body." Hence the form of the mixture is really nothing but the "broken" elementary forms themselves, in

62. Sennert, *Epitome* (1600), *Disputatio XV, Theses V* and *VI.* Cf. Zabarella, *De communi rerum*, columns 401–402.

63. Sennert, *Epitome* (1600), *Disputatio XIV, Thesis XIX:* "Nos, caeteris rejectis, Avenrois sequemur, qui putavit, non solum Qualitates, sed ipsas etiam formas Elementorum manere in misto, refractas tamen, ita, ut ex omnibus una forma fiat, non tanquam a termino a quo, sed tanquam ex partibus, non aliter, ut ex coloribus & saporibus extremis fiunt medii; ita, ut qui fuerunt gradus formarum Elementorum, jam fiant gradus formae misti."

64. This is not to be confused with the position of Avicenna, where it was only the elemental qualities, hot, cold, wet, and dry, and not the elemental forms themselves, that were "broken." Hence in the case of Avicenna one spoke of *qualitates fractae*, whereas in the case of Averroes one could refer to *formae fractae*.

65. Maier, *An der Grenze*, pp. 28–31.

combination. There is no "superadded" form of the mixture beyond the elementary forms.[66]

The striking thing about Sennert's position in 1600, however, is that he upheld the resolution up to the prime matter for all cases of corruption or mixture where a soul was not involved, in other words, wherever the corrupted thing was inanimate. Yet in the atomistic *De chymicorum* of 1619 Sennert would reject not only this position but also the less drastic one that mixture necessarily involved a resolution up to the four elements. There he specifically asks, "*[W]hether in the generation of things there must always be a resolution up to the elements*, so that there can be no true mixture except among the elements," and involving only their qualities, hot, cold, wet, and dry.[67] In response, he would demonstrate both the resolution up to the prime matter and to the four elements to be false by means of the reduction to the pristine state. This position would put him at odds both with those who held the pluralist position of Zabarella and with those followers of Aquinas—especially among the Jesuits—who maintained that there could be but one substantial form in a given substance. It is therefore clear that Sennert's turn to atomism cannot have been an outgrowth of Zabarella's approach to the plurality of forms, as at least one recent scholar has wrongly maintained.[68] In fact, it is Zabarella's theory that provides Sennert's most immediate target in the *De chymicorum*, for Sennert's restatement of his position there is effectively a recantation of his own earlier Zabarellan view. [69]

66. See Maier, *An der Grenze*, pp. 68–69 on the issue of the *forma superaddita* in Zabarella and other sixteenth-century Averroists.

67. Sennert, *De chymicorum* (1619), p. 354: "Alterum deinde quod perpendendum, est, *an semper fiat resolutio in rerum generatione usque ad elementa*, ita ut nulla vera mistio fieri posset, nisi inter Elementa, & an nulla, acidi, salsi, amari, & similium sit mistio; & porro nihil aliud alterari, & proinde ad mistionem concurrere possit, nisi quatenus calidum, frigidum, humidum, siccum."

68. I refer to Michael, "Daniel Sennert on Matter and Form," especially pp. 280–284 for Zabarella. Michael seems to claim that Sennert did not uphold the reality of *minima inexistentia*—component particles with real existence—in his chemical writings. On p. 290 he asserts the following: "In Sennert's 1611 *Institutiones Medicinae* and in his other writings on chemistry, he could maintain, consistently with Zabarella's views, that inanimate *substances* (i.e., each body that has a single form) are homogeneous bodies that, in interacting, just happen to disintegrate into particles. From the Zabarellan viewpoint, when interacting elements are transformed to produce a single new substance, the elemental particles do not endure. But particles could actually remain (as Sennert claims they do) in mixtures that are composed of several compound substances, such as medicinal mixtures." Michael's interpretation does not conform to Sennert's own assertion of atomism in the *De chymicorum* of 1619, nor does it account for his explicit rejection of the Zabarellan *resolutio ad materiam primam* there.

69. In order to drive this point home further, we may note that Sennert's rhetorical question, "[Whether] there can be no true mixture except among elements," restates Zabarella's

THE DEMONSTRATIVE POWER OF THE REDUCTION
TO THE PRISTINE STATE

How then was Sennert's reduction to the pristine state supposed to undermine the arguments of those who espoused a resolution to the elements (or further) during mixture? Sennert's 1619 experiment consists
first of fusing gold and silver together to produce a seemingly homogeneous alloy. The silver is then dissolved out by means of aqua fortis
(nitric acid), whereupon it is no longer visible in the solution. Finally,
the silver is precipitated into a "heap of innumerable atoms" by means
of salt of tartar (potassium carbonate, a base). What sort of difficulties
would this demonstration present to a scholastic who upheld Aristotelian
mixis? There are two points in the demonstration where a scholastic author would have to decide whether a homogeneous mixture or a mere
juxtaposition (a *mixtio ad sensum*) had been produced. First, the alloying
of the gold and silver led to a body that appeared homogeneous—was
this a real mixture? Second, when the silver dissolved in the nitric acid
and seemed to disappear in it, did this involve a genuine mixture of the
silver and the acid? In either case, the scholastic opponent would find it
difficult to admit that a real mixture had taken place. In the first instance,
the separation of the silver from the gold by dissolution of the former in
the acid would mean that the forms of the gold and silver had remained
intact despite their apparent combination. In the second case, the reacquisition of the silver from the corrosive acid would imply that the acid
and the silver had also retained their forms during their "mixture." This
was not supposed to be the case in *mixis*, for both the champions of the
plurality of forms theory and those who upheld a formal unity agreed
that the process leading to genuine mixture had to take place between
the elements or the elemental qualities, not between ingredients having
higher forms with their own robust being.[70]

---

claim that "elementa sola sunt per se miscibilia" (only elements are per se miscible), only to
reject the Zabarellan position. See Zabarella, *Liber de mistione*, in Zabarella, *De rebus naturalibus*,
column 478 C–D.

70. The many histories of the mixture debate that one finds in scholastic authors normally presuppose that the various representative authors, usually including Avicenna, Averroes, Thomas
Aquinas, Duns Scotus, *inter alia*, upheld a mixture between elements or elemental qualities rather
than one where higher forms combined. Good examples of such mixture histories may be found
in Zabarella, *De mistione*, columns 451–480, and Toletus, *Francisci Toleti Societatis Iesu, Nunc S.R.E.
Cardinalis Ampliss. Commentaria, una cum quaestionibus, in duos libros Aristotelis, de generatione &
corruptione, nunc denuo in lucem edita, ac diligentius emendata* (Venice: 1602, Iuntae), ff. 55r–60v.
Furthermore, as Maier points out, Scotus's mixture theory explicitly concerned the interaction

It is worth pursuing this issue further to see the tremendous difficulties that would arise in explaining how the silver was recovered upon its precipitation if one admitted that the dissolution of silver in aqua fortis was a genuine mixture. The major problem was that the silver and aqua fortis—both of them mixtures made up of the four elements—would have to lose their own substantial forms in order to acquire the substantial form of the new mixture (the *forma mixti*). Whether one argued that the process of mixture necessitated a resolution up to the prime matter or only up to the four elements, one thing was sure: a genuine mixture had to have its own *forma mixti* defining it as a true individual substance rather than a mere heap of discrete components. To the unitist follower of Thomas Aquinas, the generation of the new form of the mixture would have to presuppose the corruption of all the preceding substantial forms, so how could one then corrupt the form of the acid-silver mixture to regain the preexistent silver? After all, the silver had been destroyed during the process of mixture, even if its four elements or elementary qualities remained "virtually." To the pluralist the same problem occurred, even though the four elements remained intact.[71] In either case, once the elements were reconstituted from their virtual state (the unitist position) or the elements began to reassemble (the pluralist view), why should the elementary qualities or elements deriving from the acid-silver mixt recombine in the same proportions that they had possessed in the original ingredients, rather than forming some new substance or substances? "Reversible" chemical reactions provided both unitists and pluralists with a major difficulty—the problem of recombination.[72]

In order to make our example easier to visualize, let us imagine that a quantity of silver and acid came together to produce a genuine mixture and that some salt of tartar was then thrown into the solution with the result that the silver reemerged. Even if the silver, acid, and base each had its own characteristic proportion of fire, air, water, and earth (or

---

of the four elements *in virtute* (which did not persist numerically in the mixt) despite the fact that he upheld a version of the plurality of forms theory. See Maier, *An der Grenze*, p. 107. For more on Scotus's pluralism, see Zavalloni, *Richard de Mediavilla*, pp. 374–381.

71. Admittedly, a pluralist could have argued that the higher forms, such as the "argenteity" of the silver, could remain intact in a diminished state in order to direct the course of mixture among the elements. This would have allowed for the recapture of the silver from the mixture. Sennert's main pluralist source, however, Zabarella, did not take this position, as we have seen. Rather, he explicitly denied the continued existence of higher forms during the process of mixture.

72. See my introductory note on terminology for the restricted sense in which I use the term "reversible reaction" throughout this book.

lacking this, a fixed proportion of the qualities hot, cold, wet, and dry), the elements or elementary qualities would be free in the solution to recombine as a whole, for the forms of the silver, aqua fortis, and salt of tartar (assuming that it too entered into the process of mixture) would no longer exist to direct the ensuing recombination. Since the silver and the aqua fortis had both lost their substantial forms and become a single mixt under the new form of the mixture, why should their combined elements or qualities reassemble in the flask into new silver when salt of tartar was dropped in? Why should aardvarks or artichokes not emerge instead of a bright, heavy metal? In short, there is no reason to suppose that the dissociated four elements or elementary qualities would spontaneously recombine in the proportions necessary to form a metal instead of any other sublunary substance.[73]

If a unitist or pluralist author should, nonetheless, try to argue that the silver had been recreated anew from its elements or qualities, he would immediately fall into one of the traps that Erastus had already set for the alchemists, namely, the limitations that art placed on human creative power. No early modern Aristotelian, including Sennert, wished to argue that human beings could create substantial forms. This was the work of God and was often ranked in the same class of *creatio ex nihilo* that pertained to the Creation of the world. To create new silver out of the four elements might then be viewed as implying the creation by human means of a substantial form for the silver. If one evaded this problem by asserting that the chymist was merely imposing a preexistent substantial form on the elements, he would at once encounter another problem. Natural things, according to Aristotle, were distinguished from artificial ones by the former's innate principle of activity (*Physics* II 1 192b18–19). At the same time, however, artificial things were by definition the things that man made (*artefacta*) as opposed to God or nature. If man could create metals directly from the elements and these metals were identical to their naturally occurring exemplars, the borderline between the artificial and the natural would be seriously effaced. The creative powers of man would equal those of God and nature—he could create anything out of anything, whether "artificial" or "natural," by the imposition of substantial forms directly onto the elements. Even the alchemists did not usually maintain that their powers went this far. Geber, for example, explicitly denied that man could manufacture new metals directly from

73. A variant of this problem is described in Andrew Pyle, *Atomism and Its Critics* (Bristol: Thoemmes Press, 1995), pp. 308–309.

the metallic principles mercury and sulfur, arguing instead that the alchemist can only transform a base, but fully formed, metal, such as iron, into a more precious one. The same argument applied *a fortiori* to the artificial production of the metals directly from the four elements.[74]

There is an additional reason why Sennert's experiment presented grave problems not only for unitists, who denied the persistence of the four elements in a mixture, but also for those scholastics who might wish to uphold that some sort of enduring but altered elements, like the *formae refractae* ("broken forms") of the Averroist tradition, persisted within an aqua fortis–silver *mixis*. Unless one argued that the original silver and aqua fortis persisted in a completely unchanged state within the form of the new mixture that they composed, the *reditus* rule would emerge with all its attendant problems. As Erastus had intoned ad nauseam, there cannot be a direct return to a previous state after the removal of a form except in the case of the elements themselves; hence one substance, when it has been transmuted into another, cannot immediately be regained intact.[75] Wine, for example, once it becomes vinegar, cannot be recaptured as wine, nor can a living being be retrieved from a corpse. The only way to effect such a radical return was for the initial subject, the wine or the dead animal, to be corrupted all the way back to the elements or prime matter and then to be regenerated by following the normal circuitous course of nature. In the case of Sennert's example, this would obviously preclude the reemergence of the silver "corrupted" during its mixture with aqua fortis, since the precipitation of the silver from the acid was obviously not identical to its natural generation within the earth.[76]

But let us remove this discussion from the realm of supposition and consider a concrete and analogous example that many early modern scholastics genuinely did consider. Aristotle himself had briefly considered the case of mixing wine with water (*De generatione et corruptione* I 10 328a28–31). A genuine mixture of the ingredients could occur when the two components were roughly balanced, so that one was not merely

74. Most alchemical writers did not try to create metals *de novo* from the elements, but rather attempted to convert one already existing metal into another metal.

75. See Newman, *Gehennical Fire*, p. 152.

76. A scholastic exception was typically made for the elements, which were said to be capable of return from a mixture *in specie* but not *in numero*. What this amounted to was that a new element, say water, could be formed from water that had become air. The new water would be specifically the same as the original water but numerically different. Sennert alludes to the distinction between acquiring something *in numero* and *in specie* at *De chymicorum*, 1619, p. 686, where he is discussing extraction of alcohol from wine and the active, purging ingredient from rhubarb.

transmuted into the other, but a medial state was produced. Since Aristotle had asserted that the ingredients of a mixture could in fact be separated from one another some lines before this (I 10 327b27–29), the wine-water example could then serve as a useful illustration for discussing such reacquisition. Thus we encounter the wine-water mixture in the Jesuit Coimbrans' 1597 commentary on *De generatione et corruptione*, where they refer to Pliny's claim (*Naturalis historia*, book 16, chapter 63) that the wine would pass through a vessel made of ivy-wood, leaving the water behind.[77] But how could one maintain the reality of the *forma mixti* while also accounting for this reacquisition of the ingredients? How could one avoid the pitfall of the *reditus* principle? The Coimbrans have a ready answer, but not one that will please many modern readers: if the liquid that separates from the wine is real water, then it cannot have undergone genuine mixture with the wine in the first place. Either the wine was too weak to exercise its power on the water or there was insufficient time to do so before the separation took place. On the other hand, if the wine was sufficiently strong and there was enough time for mixture to take place, but a separation was made all the same, then "that which is extracted from the wine is not water, but a liquor of the wine itself, similar to water."[78] In other words, the Coimbrans do not see any way that genuinely mixed wine and water can again be separated.

The head-on approach of the Coimbrans was not the only one available to early modern scholastics, however. Another prominent Jesuit, Franciscus Toletus, took a markedly different tack when addressing the problem of regaining mixed ingredients. In his 1575 commentary to *De generatione et corruptione*, Toletus accepts the fact that one can recapture the ingredients of a mixture, but only in a very special sense. First he cites the testimony of the ancient commentator Philoponos, who argued that the ingredients can be separated and that "they return to their pristine

77. I have used a slightly later edition of this famous text: *Commentarii Collegii Conimbricensis Societatis Iesu, in duos libros de generatione et corruptione Aristotelis* (Lyon: Horatius Cardon., 1606), book 1, chapter 10, question 1, article 1, p. 361.

78. Conimbricenses, *De generatione et corruptione*, book 1, chapter 10, question 1, article 2, p. 363: "Quapropter quod aiunt posse aquam a vino excerni, dicendum si id quod separatur, revera aqua sit, eam excretionem posse fieri, cum vinum ita est imbecillum, ut nequeat convertere in se aquam; vel & ambo ita se invicem habent, ut ex ipsis non resultet tertium. Vel cum vinum, etsi possit in se convertere aquam; nondum tamen ei per tempus licuit eam in se convertere. Ubi vero sese istiusmodi circunstantiae non interponunt, asserendum est id, quod a vino eximitur non esse aquam, sed liquorem ipsius vini aquae similem, nondum videlicet exquisite decoctum, qui etiam ex puro vino elici potest; ut annotavit D. Thomas in 4. Dist. II. in expositione litera[e]."

degree, as when water is extracted from infused wine by means of art."[79] To Philoponos's view Toletus then contrasts that of "many Latins," meaning above all the Thomistic position: "But those who say that the elements do not remain [in a mixture] except according to their virtues say that they can be separated, because the elements that were contained in the mixture *in virtute* can again be generated out of the mixture."[80] Toletus has clearly shifted the discussion away from the problem of separating water from wine and has placed it on the very different issue of recapturing the four elements. According to the majority Latin view, he says, the elements can indeed be "separated" but only by being regenerated. A few folios later, Toletus affirms this position as his own in the following words—"I say that the elements of the mixt are produced de novo during [its] resolution."[81] This statement was relatively unproblematic, since Aristotle himself had argued that the elements can undergo a "circular" generation, whereby water, for example, becomes air and then the air becomes water.[82] But does it follow that the wine can be separated from its mixture with water by a similar regeneration? It is very hard to see how Toletus could have responded affirmatively, since he accepted the Thomistic view that all the forms of the ingredients were destroyed during the process of mixture, with only the "virtues" of the elements remaining.[83] Once the higher form of the wine had disappeared, how

79. *Francisci Toleti Societatis Iesu, Nunc S.R.E. Cardinalis Ampliss. Commentaria, Una cum Quaestionibus, In Duos Libros Aristotelis, De Generatione & Corruptione* (Venice, Juntae, 1602), fol. 54r: "Explicat Philop. quod possunt separari miscibilia, quae aliquando separantur, & redeunt ad pristinum gradum, ut cum arte extrahitur aqua, vino infuso."

80. Toletus, *De generatione et corruptione*, fol. 54r: "At vero qui dicunt non manere elementa nisi secundum virtutes, dicunt posse separari, quia ex mixto rursus elementa generari possunt, quae virtute continebantur in mixto."

81. Toletus, *De generatione et corruptione*, fol. 59r: "Ad septimum dico, quod elementa mixti resolutione de novo producuntur, ut alibi dicemus, & ex parte iam diximus." See also his comments on the preceding folio (58v): "Ad id autem, quod dicitur ipsa esse separabilia, dico primo secundum Philop. quod sensus est, non quod a mixto separentur, sed quod talia sunt ex se, ut quantum est ex se, non repugnet ipsa separari. Dico etiam, quod separabilia sunt, eo quod ex mixto possunt elementa generari rursus, sicut ex elementis generatum est mixtum, quae videlicet erant in virtute forma mixti, & fiunt in actu proprio: erant inquam in actu alieno, & fiunt in actu proprio: erant in actu formae mixti eminenter, & fiunt in actu specifico proprio, & hoc dicitur separari, ut quae erant simul in virtute coniuncta fiant seorsum seiuncta in actu."

82. See Toletus's discussion of this "circular generation" at his *De generatione et corruptione*, fol. 93v.

83. Toletus, *De generatione et corruptione*, fol. 60v. "Ex omnibus his colligo elementa non manere secundum formas proprias: ac propterea non manere actu in mixto, manere autem secundum materias, & virtutes temperatas, & loco illarum formarum elementorum succedere formam mixti, virtutes illas eminenter complectentem, atque ita manere virtute, & potentia."

could the wine itself be regained from the mixture? Despite his finessing of the issue, Toletus's position was not markedly different from that of the Coimbrans.

The position taken by Zabarella on the separation of the elements in his *De mistione* is not terribly distinct from that of the Coimbrans and Toletus, despite the fact that the Paduan philosopher upheld a version of Averroes's mixture theory rather than that of Thomas. Like Toletus, Zabarella avoids discussing the fate of the wine after the resolution of the mixture takes place. In fact, when he considers the position of Philoponos and the other ancient commentator Alexander of Aphrodisias on the separation of ingredients, Zabarella avoids giving any concrete example at all, despite the fact that these authors spoke of making experiments in separation with sponges, river-lettuce, and so forth. Instead, Zabarella shifts the whole discussion to the ethereal realm of the four elements in a way that is quite reminiscent of Toletus. Zabarella does not of course think that the elements are wholly destroyed in the process of mixture, as Thomas did. Instead, his Averroistic sympathies lead him to view mixture as the "breaking" or "dulling" of the elements up to a certain medial state, whereupon they may be said to constitute the form of the mixture. Even so, Zabarella claims to agree with the Greek commentators in saying that there can be no return of the elements "in number" (*in numero*). An element that went into the mixture—air, for example—is not the same air that comes out again. Although the initial and final air are specifically the same (identical *in specie*), the changes that the air underwent during the process of mixture were so great that it lost its full substantial identity as air. Hence it undergoes a partial regeneration (*dimidiata generatio*) upon its separation.[84] Now since an element cannot

---

84. Zabarella's distinction between regeneration *in numero* and *in specie* ultimately depends on Aristotle's discussion of "circular" versus "rectilinear" generation in the concluding chapters of *De generatione et corruptione*, book II, as at 338b6–19. See Zabarella, *De mistione*, column 471: "Ex his colligere possumus, non eandem esse hanc separationem elementorum ex misto, atque illam, qua ex mera congregatione dicuntur aliqua separari absque ulla mutatione, ut quando ex acervo tritici grana separantur: haec enim eadem numero sunt, facta separatione, quae prius in compositione, & ante compositionem erant, quia nulla est facta naturarum mutatio; & elementa separantur a misto, in quo non omnino erant actu, proinde separantur cum alteratione, quae non est absolute in accidentibus, sed est dimidiata quaedam generatio, talem enim alterationem Aristotel. in definitione mistionis intellexit; non enim fieret naturarum unio, & plurium reductio ad naturam unam, si servatis penitus formis substantialibus in solis accidentibus mutatio fieret; non possunt igitur elementa ex misto ita separari, ut redeant eadem numero, quae in mistionem venerunt." For some additional useful information regarding the medieval and early modern use of the *in numero* / *in specie* distinction, see Kangro, *Joachim Jungius' Experimente*, p. 155, n. 241.

return from a mixture identical in number to the element that entered it, clearly the same must be true a fortiori for the higher *forma mixti* of the wine, for that entity is merely a concatenation of weakened elementary forms. If Zabarella wanted to uphold the separation of wine from its mixture with water, he would have to argue for its regeneration, and here he would be confronted with precisely the same problems as Toletus and the Coimbrans.

The intractable problems that these early modern Aristotelian commentators had to face (or avoid facing) when they considered the separation of concrete ingredients from a mixt are illustrated in other contexts as well. Another example of such putative separation was the burning of green wood, which presented problems for anyone who wished both to see an elemental disaggregation in the combustion and to view the wood as a mixt in which the four elements had undergone mixture. The early modern *De generatione et corruptione* commentary falsely ascribed to Aegidius Romanus, explicitly addresses this problem, denying that the burning wood is resolved into the genuine four elements. Rather the smoke, fire, ash, and fluid are "certain imperfect mixts," which did not preexist in the unburnt wood but were manufactured by the process of combustion. On the other hand, the liquid that visibly exudes from the ends of the logs during their combustion may actually have preexisted there, but that is not relevant to the question, since Pseudo-Aegidius obviously considers this material to have been an alien substance lodged in the log rather than a component of the mixture making up the wood. The seeming elements observed during and after combustion of the wood are not the real elements but "mixts similar to the elements themselves," since the process of mixture has eliminated the possibility of a return of the genuine elements. The reason why these mixts look like the elements is that the elemental virtues remaining when the *forma mixti* of the wood was made exercise their powers on the matter during its combustion to produce them.[85]

---

85. Pseudo-Aegidius Romanus, *Commentationes physicae et metaphysicae* (Ursel: Jonas Rhosius, 1604), p. 490: "Ad quintam dicendum est illa in quae lignum viride resolvitur, non esse quatuor elementa, sed mixta quaedam imperfecta, eaque non praeexistisse actu in ligno, nisi fortasse illum humorem qui per extremitatem egreditur, nam cum in mistis sint virtutes elementorum, fit ut saepenumero dissolutio fiat in mixta similia ipsis elementis; & hoc est quod dicit Aristoteles, elementa segregari ex mixto, quanquam etiam nullum sit absurdum ex mixto generari, modo unum, modo plura elementa, quae tamen antea ibi non erant actu." The same position is upheld by the Coimbrans, for which see Conimbricenses, *De generatione et corruptione*, book 1, chapter 10, question 3, article 3, pp. 373–374. Sennert would attack this position explicitly in his *Hypomnemata*, 1636, pp. 127–128.

The position taken by Pseudo-Aegidius on burning wood contains an unintended irony, one that we have met already in the antialchemical work of Thomas Erastus. Like Erastus, Pseudo-Aegidius views the products of fire analysis as either artifacts of the fire or as heterogeneous components coexisting all along with the mixture. Let us now return to Sennert's example of the reduction to the pristine state of a gold-silver alloy in order to see how Pseudo-Aegidius would treat it. Since the forms of the metals had been destroyed during the mixing process, Pseudo-Aegidius might be tempted first to argue that the metals had been generated de novo out of the mixture in the same way that his "mixts similar to the elements themselves" were made. But here he would encounter a roadblock. Clearly he could not view the return of the intact metals as a mock formation in the way that the "mixts similar to the elements" were. The metals bear no resemblance to fire, air, water, or earth and could not be regarded as imperfect analogues of the individual elements. This left the other alternative, that the gold and silver existed together as heterogeneities in a compound, and that their substantial forms had not really been supplanted by a *forma mixti*. To admit this, however, was to play the very game that Sennert wanted, since it was equivalent to saying that the gold and silver were composed of semipermanent particles having their own unchanged substantial forms. The seeming mixture of the two metals and the subsequent mixture of the silver with the nitric acid were mere mixtures *ad sensum*—corpuscular juxtapositions rather than real mixtures, which had only fooled the eye of the beholder.

We see, then, that Pseudo-Aegidius or any other scholastic who wished to interpret the retrieval of Sennert's gold and silver as a mere *mixtio ad sensum* would be acquiescing to the corpuscular interpretation in the very act of explaining it away. But there was a further problem here as well. Could one really view Sennert's gold-silver alloy or the solution of silver in nitric acid as a mere Aristotelian juxtaposition? The cohesion of the alloyed gold and silver or the rapid evolution of gas and heating when a metal dissolves in acid is obviously something of a different order from mixing wheat and barley in a jar. In cases of *synthesis* as opposed to *mixis*, the ingredients were said to form a mere heap (*acervus*) which did not cohere. The four qualities of the elements, hot, cold, wet, and dry, were not supposed to act powerfully upon one another in such cases—that was how a real mixture, rather than a heap, came into existence. One could perhaps have attempted to escape this problem by arguing that the gold-silver alloy and the silver-nitric acid solution were mere "imperfect mixtures," which were unable to attain a complete state of *mixis*, although they had begun to coalesce. Aristotle himself had explained the formation of bronze from copper and tin in this way (*De generatione et corruptione* I 10 328b6–13) in order to account for the peculiar fact that the volume of the alloy was less than the sum of the respective ingredients. He asserted that the tin practically disappeared in the mixture, only giving its color to the copper. Yet from a scholastic perspective this could hardly account for the recapturing of the gold and silver from Sennert's alloy, since—as Toletus put it—the matter of the tin seemed to have "evaporated" during the process of mixture, leaving only its form—and hence its color—with the copper. If the matter of the silver had likewise evaporated from the gold-silver alloy, it could not have been regained intact by the mere act of precipitation from its solution.[86]

In summary, one can see how Sennert's use of dissolution and precipitation left little alternative to the conclusion that silver alloyed with gold and then dissolved in nitric acid retained its substance intact. The problems of elementary recombination, the human inability to create in the manner of God, and the *reditus* rule all led to the same conclusion— that "reversible reactions" involved corpuscular interactions rather than Aristotle's perfect mixture. But the tight bonding between immutable particles—the Geberian *fortissima compositio*—was a notion to which Sennert's scholastic opponents could not appeal: for them, the dissolution of a metal in an acid must be either a mere *mixtio ad sensum* or a mixture proper. The phenomena dictated that they could not choose the former alternative, but the latter was equally problematic. Since this phenomenon could be explained neither as an ordinary *mixtio ad sensum* nor as a proper mixture, Sennert argued that the silver was composed of minute, cohering particles, each of which retained the substantial form of silver within itself. As he was aware, the mere dissolution of salts in water or wine led to their disappearance from sight, which would then

---

86. Toletus, *De generatione et corruptione*, book 1, chapter 10, question 16, fol. 55r: "Ulterius comparat, dicitque quod unum videtur habere locum materiae, ut aes: alterum formae, ut stannum, quia non videtur misceri cum aere, sed solum colorem quendam imprimere, quod fit ex mixtione imperfecta, secundum quam parum substantiae relinquit in aere, reliquum evaporatur." The Coimbrans do not consider Aristotle's treatment of bronze in their commentary on *De generatione et corruptione*. Various other products of the sublunary world, ranging from comets to snow, were also viewed as imperfect mixtures. These types of imperfect mixture were impermanent, however, in contradistinction to the example of bronze. See Zabarella, *De naturalis scientiae constitutione*, in his Zabarella, *De rebus naturalibus*, columns 67–75.

be restored upon the evaporation of the liquid.[87] In like fashion, Sennert could argue that the optical disappearance of silver within a nitric acid solution was simply due to the fact that it had been broken into extremely fine particles: its external structure had altered, but its "internal form" remained intact. One need not assume, as previous scholastics had done, that there had been a corruption and replacement of substantial forms.

Sennert's explanation of the apparent disappearance of silver in aqua fortis involved the necessary assumption that the particles into which it was divided were extremely small. Apparently he did not think for long that the fineness of the precipitated calx was sufficient to demonstrate this, so he added an additional demonstration that appeared in his famous *Hypomnemata physica* published in 1636, the year before his death.[88] It was adjoined to the example that we have been considering as follows:

> Thus although the water in which a metal is dissolved seems to be nothing but clear water, and may be so exactly mixed that such water can be poured through paper, nonetheless the metal preserves its own unchanged nature in it and is easily precipitated at the bottom [of the flask] in the form of a very fine powder, which may then be reduced again into the metal. Thus also if a single mass be made by fusion of gold and silver together, and they come together through the smallest atoms (*per minimas atomos*) to the degree that no one may recognize that this body consists of different components, still each metal retains its own form in those minimal atoms, and can be separated by aqua fortis and can be reduced into its original body.[89]

87. Sennert, *De chymicorum* 1619, p. 362: "Ita in lixivio & muria, e qua sal, nitrum, & vitriolum coquitur; in urina, in vino sal, qui inest, non conspicitur: at separato humore, facile se conspiciendum praebet."

88. Sennert was already using the filter paper demonstration in the second edition of the *De chymicorum*. See *De chymicorum*, 1629, p. 393: "Cum enim spiritus salsi & sales, dum corpora metallorum solvunt, iis per minima permisceantur, & admistione spirituum salsorum ac salium corpora haec in minimas atomos redigantur, ita quidem ut metalla in aquis fortibus & regiis soluta per chartam transeant: sale illo alieno separato suae pristinae formae restituuntur; & licet in forma pulvis relinquantur, igne fusorio tamen pristinam formam facile acquirunt."

89. Sennert, *Hypomnemata* (1636), pp. 109–110: ""Ita quamvis aqua, in qua metallum solutum est, non nisi limpida aqua esse videatur, & tam exacte sit mista, ut talis aqua etiam per chartam transfundi possit: tamen metallum suam naturam in ea integram servat, & facili negotio forma subtilissimi pulveris ad fundum praecipitatur, qui postmodum in metallum iterum funditur. Ita etiam si una massa ex auro & argento fiat per fusionem, & ita per minimas atomos coeant, ut corpus istud ex variis constare nemo agnoscere possit: interim in minimis illis atomis quodque suam formam retinet, & per aquam fortem separari, & in pristinum corpus reduci potest. Hinc multarum operationum Chymicarum, & eorum, quae in chymicis fiunt, caussae reddi possunt."

Here Sennert added the additional step of pouring the solution of silver and aqua fortis through filter paper. The fact that no residue was left behind demonstrated the extreme minuteness of the silver particles in combination with the acid, since they had to be small enough to pass unimpeded through the pores in the paper. At this point, then, Sennert had full experimental evidence for the two theses that he had wanted to prove—that a metal seemingly mixed with an acid or another metal retained its own nature intact, and that it was composed of extremely tiny corpuscles. In doing this, he had simultaneously shown the inadequacy of the current scholastic theories of mixture while also providing a convincing demonstration of the reality of semipermanent atoms that experience no substantial modification. In this sense, one may view Sennert's use of dissolution and precipitation as an *experimentum crucis*. Although Sennert's work did not rely on Francis Bacon, his experiment provided a decisive means of picking between the two alternatives of mixture as a homogeneous mutation of matter and mixture as the association of relatively immutable particles.

FURTHER IMPLICATIONS OF THE REDUCTION
TO THE PRISTINE STATE

Needless to say, Sennert did not wish merely to discuss the single case of metals dissolved in acids and then reduced. If the apparent mixtures of a gold-silver alloy and of silver dissolved in acid were not really homogeneous, why not argue that other seeming mixtures were also really made up of heterogeneous corpuscles? This indeed was the position that Sennert took already in his *De chymicorum* of 1619. For reductions to the pristine state are not restricted, of course, to the case of metals dissolved in strong acids and subsequently recaptured. Mercury, for example, whether it be precipitated, sublimed, turned into an oil, or converted to a powder, could always be "revived" into its pristine state.[90] By the same logic that drove Sennert's reductions of metal from acid, these alterations, too, provided convincing evidence that the subject of the change underwent no substantial modification. In all such cases, Sennert argued, the substantial form of the substance that entered combination with another and was then reduced, experienced no real change at all. In the *De chymicorum*, substantial forms were divine and immutable principles imparted by God during the Creation, which determined the

90. Sennert, *De chymicorum* (1619), p. 363.

actions and passions of the substances in which they inhered.[91] Although Sennert seems to have believed, along with the Thomists and Averroists whom we have discussed, that a substantial form could be dissociated from its matter in some cases and replaced with another form, this did not mean that the form itself could undergo a change in terms of its substance-imparting characteristics.[92] More importantly, even if forms could successively replace one another, Sennert's demonstrations played on such factors as the *reditus* principle and the unlikelihood of the return of ingredients from a mixture after the loss of the *forma mixti* to undermine any claim that a substantial form had been removed and replaced during mixture. Hence the mutations in color and other properties that recoverable substances underwent had to be due to "external" factors rather than representing a change in their substantial form—"Things that are brought back to [their] minimal corpuscles and atoms put on a varying external appearance according to the varying mode of [their] combination (*pro vario concretionis modo*), even if they do not change in their internal form."[93]

We have now seen how Sennert's atomism developed through the 1610s, largely as a result of his increasing immersion in chymistry, abetted by his growing appreciation of the mixture theory propounded by Julius Caesar Scaliger. During this period, Sennert's growing commitment to atomism went hand in hand with his employment of the reduction to the pristine state as an effective weapon against the edifice of contemporary scholastic matter theory in its continuist versions. But the very fact that the reduction to the pristine state was supposed to involve no change in the substantial form of the atoms that were separating and recombining led to an obvious problem. How do we explain the

91. Sennert, *De chymicorum* (1619), p. 353: "Formae enim sunt principium divinum & immutabile, quod determinat omnes actiones & passiones rei naturalis; & sunt quasi instrumentum ac manus sapientissimi Creatoris ac Opificis Dei."

92. See, for example, Sennert's discussion of *chrysopoeia*, in *De chymicorum* (1619), p. 24: "Forma quidem ferri in formam cupri,& plumbi in aurum non vertitur; sed forma ferri decedente forma cupri, & forma plumbi decedente auri forma introducitur."

93. Sennert, *De chymicorum* (1619), p. 363: "Res autem in corpuscula minima et atomos redactae, pro vario concretionis modo, variam externam speciem induunt; etsi forma interna, non different." Sennert uses the term *concretio* to designate corpuscular combination in other passages as well. See, for example, *De chymicorum*, 1619, pp. 359–360: "Et cum [Democritus] dicit, concretione istorum corpusculorum generationem fieri, non negat mistionem, sed saltem hoc vult, vel non penetrare se Elementa, vel in mistione non semper ad elementa & materiam primam usque recurrendum esse, sed ex corpusculis antea mistis & in sua essentia constitutis nova mista generari posse."

obvious changes in color, solubility, taste, and other phenomena occurring during such reactions as dissolution and precipitation without
appealing to a change in substantial form? As we have just seen, the
German academician introduced a notion of change in "external appearance" as opposed to change in the substantial form, apparently as
a means of dealing with this problem. Sennert's claim that external appearance could be explained as a mere mode of corpuscular combination
raises the obvious questions to which we must next turn. What sort of
work did Sennert's corpuscles really do for him other than preserving
the identity of a substance in various types of change? Can we speak of
a type of reductionism in Sennert's natural philosophy? To what degree,
if any, did he invoke the structural properties of his atoms—their size,
shape, and mutual arrangement—the characteristic tools of the mechanical philosophy, for the natural and artificial processes that he chose to
describe?[94] All of these questions demand our attention, since Sennert
openly invoked the tools of the ancient atomists, *diakrisis* and *synkrisis*,
in his attempt to explain the changing qualities of matter. How exactly
did he mean to use these weapons drawn from the Democritean arsenal?
Were they mere epiphenomena arising from an obsessive desire to preserve substantial forms intact, or did Sennert make a genuine attempt to
use *diakrisis* and *synkrisis* in expanding the role of structural explanation
in chymistry? Since these questions require further probing in the Sennertian corpus before they can receive an answer, we will have to leave
their resolution for the following chapter.

94. For "structural explanation," see Ernan McMullin, "Structural Explanation," *American Philo-sophical Quarterly* 15 (1978): 139–147.

5

The Interplay of Structure and Essence in Sennert's Corpuscular Theory

EMPIRICISM AND REDUCTIONISM

The rough outlines of Daniel Sennert's early atomic theory and its development have now begun to emerge from their scholastic backdrop. It is time, then, to consider the working details of his system with a greater focus on the particularities of the laboratory. Previous writers on Sennert have, for the most part, paid little attention to the details of his corpuscular explanations, focusing rather on the theoretical underpinnings of his atomism.[1] This rather misses the point, however, for

---

1. Among these authors I include Lasswitz, *Geschichte der Atomistik*, vol. 1, pp. 436–454; Rembert Ramsauer, *Die Atomistik des Daniel Sennert: als Ansatz zu einer deutschartig-schauenden Naturforschung und Theorie der Materie im 17. Jahrhundert* (Braunschweig: Vieweg, 1935); W. Subow, "Zur Geschichte des Kampfes zwischen dem Atomismus und dem Aristotelismus im 17. Jahrhundert (Minima naturalia und Mixtio)," in *Sowjetische Beiträge zur Geschichte der Naturwissenschaften*, ed. Gerhard Harig (Berlin, 1960), pp. 161–191; Tullio Gregory, "Studi sull'atomismo del seicento II," *Giornale critico della filosofia italiana* 45 (1966): 44–63; Andreas van Melsen, *Atom Gestern und Heute* (Freiburg, 1957); Michael, "Sennert's Sea Change," pp. 331–362, idem, "Daniel Sennert on Matter and Form," pp. 272–299; and Pyle, *Atomism and Its Critics*. Kangro and Meinel are more satisfactory in that they look at the empirical details, but I disagree with their stress on the inadequacy of chymical experiment to solve theoretical puzzles about the structure of matter. See Kangro, "Erklärungswert und Schwierigkeiten der Atomhypothese," pp. 14–36; and Meinel, "Early Seventeenth-Century Atomism," pp. 68–103. The recent article by Michael Stolberg, "Particles of the Soul: The Medical and Lutheran Context of

Sennert's work was in fact deeply informed by practice and observation. For example, he adopted and refined the "negative–empirical" approach of Geber in considering the limits of technical analysis to provide the natural philosopher with a working "atom," namely, any substance that was resistant to dissolution in the laboratory. This important idea is already clearly stated in the 1619 *De chymicorum* and repeated in the later *Hypomnemata physica*. Sennert says that the three principles of the Paracelsians, mercury, sulfur, and salt, are principles in the sense that "art can hardly progress further [than these] in the resolution of natural things, nor perhaps may even nature proceed, who when she constitutes something as a mixt, constitutes it from these *prima mixta* rather than immediately from the ultimate simples."[2] A *primum mixtum* (first mixt), then, was a substance presumed to be composed of the four elements: in practice, however, it could not be further analyzed.[3] The fact that such semipermanent substances had a particulate structure was revealed to the eyes by the quotidian operations of chymistry. Sublimation, for example, separated a substance into its "minimal corpuscles," and these "atoms" then collected on the internal walls of the alembic or still.[4] And as we saw in the case of Sennert's 1619 reduction of silver from nitric acid, he believed that the senses could provide direct access to the "congeries of

---

Daniel Sennert's Atomism," in *Medicina nei secoli* 15 (2003): 177–203, makes a promising foray into the medical and theological dimensions of Sennert's atomism, but Stolberg also falls victim to the easy habit of downplaying Sennert's empirical evidence without seriously considering the scholastic theories that Sennert intended his work to overthrow.

2. Sennert, *Hypomnemata* (1636), p. 41: "Hoc loco ut pauca de iis dicam, primo etiam ii, qui ea simpliciter non admittunt, prima mixta esse concedant. De quo cum nemine litigabo, modo hoc obtineam, eo modo principia dici posse, quod in resolutione rerum naturalium ars ultra progredi vix possit, imo nec natura forsan progrediatur, dum aliquod mistum constituit, illud non immediate ex ultimis simplicibus, sed potius ex primis istis mistis constituit." The same idea is found in the *De chymicorum* (1619), p. 282, but without the term *prima mixta*.

3. The important criterion that a *primum mixtum* atom is operationally indivisible is further elaborated in the 1636 *Hypomnemata physica* (pp. 107–108): "Sunt enim secundo alterius, praeter elementares, generis atomi, (quas si quis prima mista appellare velit, suo sensu utatur), in quae, ut similaria, alia corpora composita resoluuntur. Et omnino in mistione rerum naturalium, seu quae fit in non viventibus, corpora, e quibus mista constant, ita in exiguas partes confringuntur, & comminuuntur, ut nullum seorsim, & per se agnosci possit. In omnibus etiam fermentationibus & digestionibus ac coctionibus, quae vel a natura, vel ab arte fiunt, nihil aliud agitur, quam ut ad minima redigantur, & ea sibi arctissime uniantur. Contra resolutio corporum naturalium, cum ea, quae a natura, tum quae arte fit, nihil aliud est, quam in minima corpora resolutio."

4. Sennert, *De chymicorum* (1619), p. 361: "Idem operationes Chymicae testantur; imprimis sublimatio, ubi Atomi illae in Alembico colliguntur."

atoms" found in the form of a powdery precipitate at the bottom of the flask.[5]

The idea that one could actually see atoms may sound terribly naïve to the modern reader, but it is necessary to stress that Sennert used the word "atom" in very much the same way that his predecessor Geber had employed the term "part" (*pars*). An "atom" could refer either to the bits making up the four elements, the corpuscles of mercury, sulfur, and salt made up of the elements, or the particles of a metal or some other higher substance composed of the principles. In the late *Hypomnemata physica*, Sennert would therefore say that there were different *genera* of atoms and that these reflected different stages of corpuscular composition.[6] When Sennert speaks of seeing atoms in a flask or sublimatory, he is referring to what he takes to be atoms of higher compositional stage, not the tiny elemental corpuscles out of which the former are composed. Nor does he mean that the chymist can observe an individual atom distinct from the particles with which it is associated. Rather, the macrolevel observation that sulfur sublimes as "flowers" or that silver reduces as a powder reveals a discontinuous structure instead of an unbroken continuum.

The empirical character of Sennert's atomism reveals itself further in his refusal to speculate on the shapes, arrangements, motions, or other imperceptible characteristics of his corpuscles. Here again he follows Geber and the long tradition of alchemical corpuscular theory. The minuteness of the bits (or in the case of mercury, mistlike droplets) that first collect on the alembic during sublimation is evident to the eye, as is the smallness of the precipitated corpuscles of silver. The residue that is left at or near the bottom of the flask when many substances are sublimed does not usually have this fine character, but exists in the form of fused lumps or charred remains.[7] Is there not, then, an obvious correlation

5. Sennert may have modified this view in his later works, for the *Hypomnemata physica* does not repeat the expression "congeries atomorum" in its parallel description of the reduction from acid. This could be due to a growing awareness on Sennert's part of the incredible minuteness of his atoms, revealed by filtration and other means.

6. Sennert, *Hypomnemata* (1636), p. 94: "Cum vero atomorum non sit unum genus, sed pro corporum naturalium varietate varia; eas & secundum simplicia corpora, quae elementa dicuntur, & secundum composita, considerare libet. Primo enim ipsa elementa in talia corpora resolvuntur, & corpuscula rursum coeuntia, tum composita corpora, tum ipsam molem elementorum constituunt," and 142: "Ideoque etiam corpora illa, ex quibus haec fiunt, etsi minima dicuntur, tamen absolute talia non sunt, sed sui generis minima, id est, talia, in quae corpora illa, cum resolvuntur, abeunt, atque ita non in elementa, sed in ea, e quibus proxime constant."

7. This observational point is made very clearly in Geber's *Summa perfectionis*. See Newman, *Pseudo-Geber*, p. 384: "Est et similiter intentio altera ut semper seorsum separetur quod sursum ad propinquitatem foraminis capitis aludel ascendit in pulverem ab eo quod fusum

between corpuscles of small size and ease of sublimation? This and similar observations led the tradition of alchemical corpuscularism from which Sennert imbibed to allow a considerable degree of speculation about the relative sizes of particles, even though one could say nothing of their shape or arrangement. Yet it was obvious that corpuscle size alone could not account for the polyvalent experiences of the laboratory, even in the case of sublimation. Many powdery substances, such as calcined limestone and salt of tartar, are highly resistant to being sublimed. Recognizing this, the chymist had two choices. He could either engage in further microstructural speculation to explain the stubborn fixity of some powdery substances, employing the elaborate hooks, eyes, and protuberances of Nicolas Lemery and various other post-Cartesian corpuscular chymists, or he could admit, as Sennert did, that there were different genera of corpuscles whose properties varied not only with their physical structure but also with their substance.[8] In Sennert's system, hylomorphic explanations based on the substantial form interacted with structural explanations partly derived from Geberian alchemy and partly based on Sennert's understanding of Democritean atomism. The result of this characteristic fusion was an approach to matter theory that was at once both reductionistic and stubbornly wedded to the view that there were irreducible qualities lodged within the corpuscles of matter, which "flowed from" the substantial forms.[9]

Since Sennert presents us with two complementary sources of qualitative difference—namely, the structural characteristics of his corpuscles and the essences residing in their substantial forms, we must attempt to disentangle the divergent realms in which these two types of agency work. We will first consider the structural operations of the corpuscles and then return to the qualities flowing from their substantial forms. As we have noted, Sennert already employed the atomistic terms *synkrisis* and *diakrisis* in his *Institutiones medicinae* of 1611 to designate the

---

et densum in frustis et apud fundum illius pervium et clarum cum adherentia vasis ad spondilia conscendisse invenitur."

8. On Lemery, see Michel Bougard, *La chimie de Nicolas Lemery* (Brepols: Turnhout, 1999).

9. Here I must raise an objection to Stolberg's blanket use of the term "anti-reductionist" for Sennert's atomism. It is true, of course, that Sennert strongly rejected the jejune reductionism of certain scholastic authors who attempted to account for complex phenomenal change in terms of the four elemental qualities. If we examine Sennert's own explanations of change, however, they often make significant appeals to structural reductionism, as I show in the present chapter. Hence we cannot simply label Sennert an "anti-reductionist" and leave the matter at that. Instead, it is imperative that we work out the details by which he integrated structural explanations with hylomorphism. For Stolberg's view, see his "Particles of the Soul," pp. 184–186 and passim.

separation and association of particles. He points out in the same text, however, that the combination and separation signified by these two terms do not entirely suffice to explain the totality of chymical operations. In addition, he says, there is a third class of change, which he calls "alteration, immutation, perfecting, and conserving."[10] In later texts, Sennert would refer to this third category merely as *immutatio* ("immutation"), a Latin term that can be defined simply as "change" or more specifically as "substitution" or "replacement," as when one word is substituted for another in a rhetorical speech. In the *Institutiones medicinae*, Sennert asserts that this "immutation" pertains especially to processes such as "digestion" and circulation in a sealed flask or still, where a change is gradually brought about in the enclosed material but without any obvious separation or aggregation. Sennert advances this idea considerably in the second edition of his *De chymicorum* (1629), which contains a weighty appendix that is lacking in the 1619 version of the text (the new appendix bears the title "De constitutione chymiae" [On the Organization of Chymistry]). Here he defines immutation as the induction of "a new mode of substance or a quality" brought on neither by a segregation of corpuscles (*diakrisis*) nor by an aggregation of them (*synkrisis*).[11]

It is likely that Sennert had in mind some sort of corpuscular rearrangement when he spoke of immutation, as his reference to the imposition of a "mode of substance or a quality" might suggest. With these words he clearly did not mean to imply substantial change itself, for in a passage that we introduced in the last chapter, Sennert explicitly said that things "that are brought back to [their] minimal corpuscles and atoms put on a varying external appearance according to the varying mode of [their] combination [*pro vario concretionis modo*], even if they do not change in their internal form."[12] The mode of corpuscular

10. Sennert, *Institutiones medicinae* (1611), p. 1046: "In tertia vero eas explicabimus, quae ad rei alterationem, immutationem, perfectionem, conservationemque sunt comparatae: quales sunt; digestio, circulatio, conditura, nutritio."

11. Sennert, *De chymicorum*, (1629), p. 394: "Tertium adhuc operationum Chymicarum genus superest, cum scilicet neque quae unum sunt solvere & segregare, neque, quae separata sunt unire & coniungere cupimus, sed rei novum substantiae modum vel qualitatem inducimus, quam operationem Immutationem appellare libet."

12. Sennert, *De chymicorum* (1619), p. 363: "Res autem in corpuscula minima et atomos redactae, pro vario concretionis modo, variam externam speciem induunt; etsi forma interna, non differant." Sennert uses the term *concretio* to designate corpuscular combination in other passages as well. See, for example, *De chymicorum* (1619), pp. 359–360: "Et cum [Democritus] dicit, concretione istorum corpusculorum generationem fieri, non negat mistionem, sed saltem hoc

combination here accounts for a mutation in external form when the internal, substantial form remains unchanged. In referring to immutation, Sennert was probably thinking of the *metakinēsis* or "transposition" resulting from changes in *thesis* and *taxis* (position and arrangement) that Aristotle claimed to account for mere alteration in the Democritean system as opposed to *synkrisis* and *diakrisis*, which were supposed to explain generation and corruption (*De gen. et corr.* 315b6–15). But again, the empirical nature of Sennert's investigation prohibited any explicit speculation as to how such corpuscular rearrangement might occur and whether his corpuscles would take on new structural characteristics. Thus instead of going down the path of a Descartes or Lémery, Sennert argues again from the phenomena. He provides a number of operations where an observable change in consistency and other properties occurs without any noticeable material loss or gain, as when metals are supposedly rendered liquid and potable per se and without addition by chymists, when a liquid is solidified without addition, such as mercury "precipitated" per se (i.e., repeatedly sublimed until it falls down as a powder), when a substance is fixed or volatilized without addition or loss, when a material is converted to a Paracelsian "magistery," which again occurs without loss or addition of matter, or when a substance is vitrified by intense heating.[13] In all such cases, Sennert says, the substance is homogeneous and nothing is separated from it, but a new mode of substance is induced by the laboratory operation.[14]

We see, then, that Sennert upheld three sorts of corpuscular activity— the Democritean association of atoms called *synkrisis*, the parallel dissociation of atoms referred to as *diakrisis*, and the vaguely described rearrangement of atoms that Sennert dubbed "immutation." Although immutation plays a quite restricted role in Sennert's explanations of phenomena, the case is otherwise with *synkrisis* and *diakrisis*, which swell the pages of his *De chymicorum*. Already in the 1619 edition of that text, Sennert accounts for the putrefaction, combustion, and vaporous

---

vult, vel non penetrare se Elementa, vel in mistione non semper ad elementa & materiam primam usque recurrendum esse, sed ex corpusculis antea mistis & in sua essentia constitutis nova mista generari posse."

13. Sennert is actually quite skeptical about the claim that metals can be rendered potable per se or that the magisteries sold by chymists really lack additaments. Nonetheless, if these products can genuinely be made per se, they belong in the genus of immutation, so Sennert discusses them in his treatment of that subject. See *De chymicorum* (1629), pp. 394–399.

14. Sennert, *De chymicorum* (1629), p. 399: "Ea tamen hic intelligimus, quae sunt plane homogenea, & a quibus nihil separatur, sed solum novus substantiae modus iis inducitur."

exhalation of all things by means of *diakrisis*. It is only an illusion of the eye that makes us think the vapors from such disintegrating bodies are continuous, for they are really composed of "many thousands of atoms, mixed together among themselves."[15] Burning pitch, for example, releases a smoke made of multitudes of atoms, a process that is similar to the sublimation of atoms in an alembic. *Synkrisis*, on the other hand, occurs when the stony material dissolved in mineral waters in the form of minimal particles congregates to form hard rocks. It also takes place when the resolved particles of food are gathered together and assembled into flesh, blood, or other matter in the process of digestion, just as the atoms come together in the artificial "digestions" that are performed in laboratory vessels.[16] The repeated elision that Sennert makes here between chymical operations performed in vessels and natural processes occurring at large is no accident, for it is precisely these humble procedures of the laboratory that supply him with the empirical evidence of atomic interaction that he then transposes onto the natural world.

If we turn, then, to the 1629 edition of Sennert's *De chymicorum*, we find all of these observations repeated, and expanded considerably in the appendix *De constitutione chymiae*. This appendix, in fact, may be described as a comprehensive discussion of laboratory operations in terms of *synkrisis*, *diakrisis*, and immutation. The new appendix explicitly clothes the corpuscular and Geberian reading of these operations that Sennert gave in his *Institutiones medicinae* of 1611 in a language that the author now acknowledges to be Democritean and atomistic. The 1629 appendix describes the dissolution of metals in acids, not surprisingly, as a *diakrisis*. Whereas some cases of *diakrisis* divide a substance into heterogeneous parts, however, dissolution in acid simply breaks the metal into homogeneous metallic atoms.[17] This division occurs because the metals, containing the Paracelsian principle salt, are attracted by the saline spirits in the acid menstruum on the principle that like goes to

15. Sennert, *De chymicorum* (1619), p. 361: "Ubi enim ex re aliqua, quae vel putrescit, vel comburitur, vel alias a calore resolvitur, vapor aut fumus attollitur, Visus quidem e longinquo corpus continuum esse judicat; Cum tamen non sit continuum, sed atomorum multa millia inter se confusa, ut vel visu intentiore animadverti potest."

16. Sennert, *De chymicorum* (1619), p. 363: "Et quid aliud sunt digestiones & coctiones, cum eae, quae arte instituantur, tum quae a natura in corporibus plantarum & animalium fiunt, quam primo διάκρισις & corporum miscendorum in minimas partes resolutio: iterumque pro rei cuiusque natura & usu σύγκρισις & concretio."

17. Sennert, *De chymicorum* (1629), p. 390.

like.[18] The saline spirits draw the metallic atoms to themselves and unite with them to form a complex corpuscle. As Sennert puts it, "[T]he metal having been united to the menstruum, or certainly to the salt that is in it, assumes another external form and is rendered either a liquid, powder, or saltlike substance, while the salt in the corrosive menstruum adheres to the metallic body."[19] Sennert's clear description of an aggregate corpuscle endowed with a new "external form" reaffirms his belief that no substantial change occurs when metals are dissolved in acids.

The importance of corpuscular aggregation in changing the perceptible properties of matter receives further elaboration in Sennert's consideration of "sugar of lead" (*saccharum saturni*) or lead acetate. Chymists made sugar of lead by reacting vinegar with metallic lead—the toxic compound received its name from its sweet taste. As Sennert was aware, one could then decompose the compound by distillation to acquire a highly volatile, burning spirit (primarily consisting of our acetone). He explains this in corpuscular terms in the *De chymicorum* of 1629 but also somewhat less guardedly in a letter written to his friend and former student Michael Döring in 1623. According to Sennert's letter, the volatile spirit does not derive from the lead itself. Rather, it comes from the acid component of the vinegar, which is itself a salt derived from the spirit of wine (impure ethyl alcohol). When the sugar of lead is subjected to distillation, the heat causes this vinous salt to disengage from the lead, hence regaining its freedom. According to Sennert, then, the volatile spirit separated from sugar of lead is spirit of wine, while spirit of wine itself is merely a more volatile form of vinegar, and by the same logic vinegar is just a fixed form of spirit of wine. "All of this depends on atoms," he assures Döring, "but how? "Spirit of wine and spirit of vinegar consist of the same material, and of the same species of atom. But there is one arrangement (*positus*) of them and one aggregation (*unio*) in spirit of wine, and another in spirit of vinegar."[20] This is perhaps the closest

18. In Sennert's day, the three strong acids, sulfuric, hydrochloric, and nitric, were all derived by distillation from substances that were acknowledged to be "saline," such as iron sulfate, table salt, and saltpeter. Hence he refers to them as "saline spirits."

19. Sennert, *De chymicorum* (1629), p. 387: "Ideoque sal vel salini spiritus (alii tamen aliis metallis magis cognati sunt) qui sunt in menstruo solvente, metalli partes atomorum quasi modo ad se trahunt sibique uniunt: atque metallum menstruo, aut certe sali, quod in eo est, unitum aliam formam externam induit, & vel in liquorem vel pulverem, aut saliformem substantiam redigitur, & sal, qui est in menstruo solvente, corpori metallico adhaerescit."

20. Sennert, *Opera omnia* (1676), vol. 6, p. 592; letter dated 23 March 1623: "Acetum, quatenus acidum, quia acre, & hanc suam vim a sale vini habere puto.... Confirmatque hanc opinionem, praeter ea, quae a te allegantur, valde illud (modo verum sit) quod Angelus Sala contra

that Sennert ever comes to linking specific chymical properties at the macrolevel to modifications in the structural properties of the corpuscles making them up. Despite his appeal to structure, the attraction that implicitly unites the two types of atom—the metallic and the acid that jointly make up the sugar of lead—is clearly due to affinity rather than any kind mechanical impulsion.

Sennert views the precipitation of a metal (or other material) from an acid as a *synkrisis* since it involves the aggregation of the homogeneous atoms that had been separated during the process of dissolution. How does this *synkrisis* come about? The particles of corrosive spirit united to the smallest atoms of the metal float in the solution "as if one body" so long as they remain bonded, for they are attracted to the remaining corrosive liquor. But if the atoms of corrosive spirit separate from the bonded metal, the metal sinks to the bottom of the flask in the form of a powder. This easily occurs if one pours in some alkaline liquid such as oil of tartar (dissolved potassium carbonate), for which the corrosive has a greater affinity than it did for the metal. In such a case, the acid and the oil of tartar bond, thus releasing the metal.[21] The reader who hopes to find a purely "mechanical" explanation in terms of the size, shape, and arrangement of Sennert's atoms will be sorely disappointed by this account. And yet we cannot fail to be struck by the closeness of Sennert's description to the elective affinities recounted by Isaac Newton in query 31 of his *Opticks*. Newton, like Sennert, employs corpuscles endowed with attractive powers to explain the association that some substances have for others, and in a similar fashion he argues that this attraction comes in varying degrees. It is precisely the fact that nitric acid has a greater affinity for oil of tartar than it does for the metals that allows the latter to separate from the acid and precipitate. Beyond this emphasis on the mutual displacement of corpuscles with varying degrees of elective affinity, Sennert's system also shares another explanatory feature with Newton's. According to Sennert's *Hypomnemata*, the heat and boiling that

---

Quercetanum defendit; ex plumbo nullum fieri spiritum, sed quod inde fieri videtur esse spiritum Vini, qui ex aceto, quo plumbum solutum est, postliminio redit, & quasi reviviscit, Nimirum dum spiritus Vini figitur, fit inde acetum; dum spiritus aceti volatilis redditur, fit inde spiritus Vini. Quae omnia ex atomis pendent. Constat enim Spiritus Vini & spiritus Aceti eiusdem materiae & speciei atomis: sed alius positus, aliaque eorum unio est in spiritu Vini, alia in spiritu Aceti. Vides non esse de nihilo, quae nuper de atomis Democriti ad te scripsi; Et saepe invito, & aliud cogitanti, necessario quasi de iis aliquid cogitandum est." The parallel passage is found in Sennert, *De chymicorum* (1629), pp. 424–425.

21. Sennert, *De chymicorum* (1629), p. 392.

often occur when a substance is dissolved in an acid can be explained as "the sudden motion of similars to similars."[22] In other words, the atoms of the dissolving body, having an affinity for the particles of aqua fortis, rush towards them at great speed, causing heat and ebullition. They are then bonded until they may be induced either to precipitate or bond with some other substance. Newton likewise saw the violent heat given off in such reactions as a product of the invisibly small motion of the corpuscles rushing toward one another.[23]

Sennert's focus on chymical affinities has its counterpart in his acceptance of the old belief that some substances have an antipathy for one another. His views on this subject are tempered, however, by his usual critical reserve. Hence he objects to the claim of the chymist Angelus Sala that mercury is precipitated from an acid solution by salt of tartar as a result of an antipathy between the two substances. Sennert insists, to the contrary, that it is the very affinity between salt of tartar and acids that forces it to bond with them and release the mercury. He goes so far as to express as a general rule that "precipitation [from a solution] universally comes about when something is dropped or poured into the solution by whose power the solvent liquid, or that which was the cause of solution in the liquid, is separated from the dissolved body."[24] Nonetheless, antipathies present themselves in different circumstances. The deflagrating and explosive effects of gunpowder, for example, are due to the antipathy that niter has for easily ignited materials such as sulfur and charcoal. The unsophisticated think that gunpowder's easy deflagration is due to the high inflammability of the niter in it. But Sennert points out that niter can be thrown on a red-hot iron plate without catching fire—instead it melts and sublimes. Gunpowder's rapid inflammation does not occur merely because the niter itself is ignited, he says, but because it flees the burning sulfur and charcoal so rapidly. Why, then, can one drop sulfur onto molten niter without causing an effect similar to that of gunpowder? Precisely because the antipathy between the two substances prevents their intermingling. The same thing can be seen in water and oil, which have an obvious mutual antipathy. At room

22. Sennert, *Hypomnemata* (1636), p. 37: "Ubi tamen hoc notandum, credibile esse, istam ebullitionem non saltem fieri ex pugna contrariorum, sed etiam subito motu similium ad similia."

23. Isaac Newton, *Opticks* (New York: Dover, 1952), "Query 31," pp. 377–378.

24. Sennert, *De chymicorum* (1629), p. 392: "Et qui omnes omnino, cujuscunque sint generis, praecipitationes perpenderit, animadvertet, praecipitationem in universum fieri, cum in solutionem aliquid injicitur vel infunditur, cujus vi liquor solvens, seu id, quod in liquore solutionis caussa fuit, a corpore soluto separatur."

temperature they calmly separate, because there is only a superficial contact between them. But if the oil is ignited and water is then poured on, the two substances are forced to mix *per minima*, whereon the inflamed oil leaps into the air in a desperate attempt to escape. This is precisely what happens with burning gunpowder, only here the substances involved are not liquid, but solid. Hence they must be ground very subtly and their particles juxtaposed in order to attain the very *mixtio per minima* that they will flee when an external flame is applied. The resulting explosion that occurs when the gunpowder is tightly constricted merely follows from the rapid expansion of the materials involved in the form of vapor.[25]

Sennert's treatment of niter shows once again the interplay between the two types of qualitative explanation in his work—the structural and the substantial. On the one hand, niter has an innate antipathy for sulfur and charcoal, no doubt due to their substantial forms. But in order for this antipathy to be realized in the deflagrating effect of gunpowder, the substances must be ground into tiny particles that have, in turn, to be moved into close proximity with one another. Only in this way will the maximum surface contact be obtained between the three ingredients, bringing their antipathy to its full realization and causing their violent expansion. It is the combination of the antipathy that flows from the substantial form and the small size and close proximity of the corpuscles that results in gunpowder's spectacular effects. Here again Sennert uses a generalized microstructural explanation in combination with chymical properties originating in the substantial form to explain the origin of a macrolevel effect. And as in the case of the so-called atoms seen by Sennert during sublimation and precipitation, the empirical bases of his comments are easily discerned. It is an indisputable fact obvious to anyone who takes the time to experiment that the ingredients of gunpowder really do have to be ground into fine particles for it to achieve an explosion. And Sennert's comments about the inflammability of its ingredients could be verified by the simplest means available in any chymical laboratory. The absence of more elaborate explanation at the microlevel was, as usual, an attempt by Sennert to adhere to the observable phenomena. Can this also be said of his appeal to the causal agency supposedly responsible for chymical properties such as attraction and repulsion, namely, the substantial form? We will now address this question.

---

25. Sennert, *De chymicorum* (1629), p. 409: "Atque hinc procul dubio etiam est pulveris pyrii vis, dum non solum sulphur & nitrum accensa locum ampliorem requirunt, sed & sese aversantur."

SUBSTANTIAL FORMS, OCCULT QUALITIES, AND THE LIMITS
OF OBSERVABILITY

Substantial forms, for Sennert, were unknowable. He was fond of quoting Avicenna and Scaliger to that effect, though they, of course, were merely expressing a widely held view in particularly trenchant fashion.[26] After all, did the Persian philosopher not say that the form of fire itself was no more accessible to human sense than the so-called occult qualities of the scholastics, such as magnetic force or the mysterious ability of poisons to infect an entire body in a trifling dose?[27] And did not Scaliger assert that form is a divine thing, an exact cognition of which escapes us?[28] It was this very principle of per se unintelligibility that would provide the mechanical philosophers of the seventeenth century with some of their most telling ammunition against scholastic thought as a whole.[29] It was a fundamental principle of scholastic Aristotelianism that to know a thing is to know it through its cause. But many scholastics themselves admitted that form, the supposed origin of qualities, was imperceptible, and even per se unknowable. How then could they deny the advantages of a system that replaced hylomorphic explanation with the easily conceivable interactions of microstructural machines? Yet there is another way to look at the principle of formal nescience, at least in the hands of Sennert.

26. Thomas Aquinas and his followers, for example, stressed the inaccessibility of the substantial form to the senses and the consequent necessity of relying on the accidents for knowledge of it. See Thomas Aquinas, *Commentarium in libros de generatione et corruptione*, in *Sancti Thomae Aquinatis doctoris angelici opera omnia* (Rome: Typographia Polyglotta, 1886), book 1, chapter 3, lectio 8, article 5, p. 293. See also Aquinas, *Commentarium in libros metaphysicorum*, book 7, lectio 2, *Quaestiones de veritate*, question 10, article 1, response to 6, and *Summa theologiae*, part 1, question 29, article 1, response to 3. For more on the substantial form and its imperceptible character, see Brian Copenhaver, "Scholastic Philosophy and Renaissance Magic in the *De Vita* of Marsilio Ficino," *Renaissance Quarterly* 37 (1984): 523–554, especially pp. 539–549.

27. This is Sennert's interpretation of Avicenna, whom he cites at *Hypomnemata* (1636), pp. 46–47. For Avicenna's assertion of the unknowability of the cause of fire's properties, see his *De viribus cordis*, in *Avicennae arabum canon medicinae* (Venice: Iunctae, 1608), vol. 2, pp. 340–341.

28. Scaliger, *Exotericarum exercitationum* exercitatio 307, section 29, quoted by Sennert, *Hypomnemata* (1636), p. 46.

29. For this reason, Boyle prominently displays quotations from Scaliger, Thomas Aquinas, and Sennert affirming the unknowability of forms on the first page of his text "Of the Origine of Forms," in *The Origine of Formes and Qualities*. See Michael Hunter and Edward B. Davis, *The Works of Robert Boyle* (London: Pickering & Chatto, 1999), vol. 5, p. 339. See also pp. 340, 351–352, of *Forms and Qualities* and also Boyle's comments in the preface to "A Physico-Chymical Essay," in Hunter and Davis, *Works*, vol. 2, p. 87, where he says that the most ingenious of the scholastics admit that substantial forms are incomprehensible.

We have seen the nature of his empiricism in the careful limits that Sennert placed on microstructural explanation. He was not one to stray far from the observational. The brute realities of elective affinity, inflammability or its absence, volatility or fixity, and a host of other phenomena could not be understood on the basis of the size of corpuscles alone, any more than these facts of the laboratory were subject to convincing explanation by means of *synkrisis, diakrisis*, and immutation. The state of having small corpuscles or the ability of the corpuscles to be separated, congregated, and rearranged could be shared by any number of substances with widely varying properties. The powdery substance quicklime was fixed, infusible, resolutely stable in fire, and reacted violently with water to produce heat. Niter, on the other hand, which could also be ground to a powder, was somewhat volatile, quite fusible, an ingredient of the explosive gunpowder, and cooling when placed in water.[30] In Sennert's corpuscular theory, these properties were not capable of complete explanation by structural factors alone—considered over a variety of substances, small particle size, for example, could be a necessary condition of volatility, liquidity, and mixture, but it could not be a sufficient one.

Having ruled out microstructural explanations as a means of arriving at a complete index of the essential differences of substances, Sennert could have adopted a scholastic strategy already employed in antiquity, arguing that the differing actions of things derived from their varying quantities of the four elements or of the four elementary qualities, hot, cold, wet, and dry.[31] But what empirical evidence could one adduce to demonstrate that the bitterness of a prune pit, the greenness of grass, or the putrid smell of rotten meat derived from a particular mixture of elements or qualities? No such demonstration was at hand, and Sennert directed much of his energy to the destruction of such impoverished attempts at elementary explanations. As we have already seen, his attack on the prevailing scholastic theories of mixture was itself an attempt to discredit the idea that profound combination required an action of the elements or their qualities on one another (this action was called by scholastic authors the *pugna elementorum*—the "fighting of the elements") that would pave the way for the imposition of a new substantial

---

30. Potassium nitrate begins to decompose at 400°C, releasing oxygen.

31. J. C. Scaliger referred to the "malus genius Alexandreus" that supposedly led Girolamo Cardano to this approach, in honor of the ancient Aristotelian commentator Alexander of Aphrodisias. See Scaliger, *Exotericarum exercitationum*, exercitatio 101, section 14, p. 284. Sennert quotes this passage at *De chymicorum* (1619), p. 171.

form. In effect, Sennert was trying to displace the elements as universal causal entities by more complex corpuscles endowed with their own substantial forms. But what evidence was there for the substantial form itself? Why should we admit that any such principle of essential identity exists? The answer to this question can be viewed, paradoxically, as an affirmation of Sennert's empiricism. The qualities that flow from the substantial form are either sensible, like redness, or produce sensible effects, like the observational fact that a piece of iron is attracted to a lodestone. The substantial form itself is completely insensible, a causal *terminus post quem*, from which perceptible qualities arise without revealing the nature of their source. Substantial form, therefore, is a sort of "black box" from which qualities emerge. The unknowable nature of Sennert's substantial form is an Aristotelian empiricist's statement of nescience.

Sennert's attitude toward substantial form as a "black box" finds a parallel in his related treatment of the so-called occult (i.e., "hidden") qualities of the scholastics. Scholastic authors who, like Sennert, viewed the attempt to reduce all the phenomena of the world to the interaction of hot, cold, wet, and dry as absurd, postulated the existence of additional but imperceptible qualities to account for change where no sensible cause was present. Galen, writing in the medical context of late antiquity, had referred to occult qualities as flowing from "the whole substance" of a thing, as opposed to deriving from its elementary qualities. Obvious instances abounded in nature—the phenomena of magnetic attraction, electrical shock by the "torpedo" (electric eel), the supposed ability of the mythical fish *echeneis* to stop a ship at sea, purging by means of drugs, the virulence of fast-acting poisons, allergy to cats, and many other cases displayed singular effects that could not receive easy explanation in terms of the tangible qualities hot, cold, wet, and dry. In scholastic parlance, such effects were manifest, while their causes were occult.[32] Sennert was deeply interested in occult qualities from his earliest days as a student at Wittenberg. His teacher, Johann Jessenius, had already set Sennert the task of orally defending his positions on the subject in a 1596 disputation concerning diseases of the "whole substance" and

---

32. On the history of occult qualities, see Copenhaver, "Scholastic Philosophy and Renaissance Magic." See also Copenhaver, "Astrology and Magic," in Charles B. Schmitt and Quentin Skinner, *The Cambridge History of Renaissance Philosophy* (Cambridge: Cambridge University Press, 1988), pp. 264–300; and Copenhaver, "The Occultist Tradition and Its Critics," in Daniel Garber and Michael Ayers, *The Cambridge History of Seventeenth-Century Philosophy* (Cambridge: Cambridge University Press, 1998), vol. 1, pp. 454–512.

in a 1599 *disquisitio* on the causes of sympathy and antipathy.[33] Among other things, Jessenius points out an important scholastic principle that seems to necessitate the existence of occult qualities—"nothing can act beyond the faculty and consequence of its own form"—the faculty of fire, for example, is to heat, so it cannot attract iron like a lodestone. Since the manifest qualities can only heat, cool, wet, and dry, they cannot be responsible for the marvelous effects that have insensible causes. The topic of occult qualities comes up again in Sennert's 1599–1600 *Epitome naturalis scientiae*, where he, like Jessenius, is inclined to accept that occult qualities come down to us from celestial bodies.[34]

Sennert also derived from Jessenius a deep respect for two medical authors who had written influential works relating to the issue of occult qualities—Jean Fernel (c. 1497–1558) and Girolamo Fracastoro (c. 1478–1553). Fernel is famous today for the strong support of occult qualities given in his *De abditis causis rerum* of 1548. He argued there that plague and various epidemics cannot be explained as merely being due to elemental or humoral properties but result from occult qualities, which in turn derive from the substantial form. Following the Neoplatonism of Marsilio Ficino, Fernel claimed that substantial form itself originates in the supramundane regions and is transmitted by the heavenly bodies to earth.[35] Fracastoro, on the other hand, published his famous *Syphilis sive*

33. The 1596 disputation bears the following title: *De morbi, quem aer tota substantia noxius peragit, praeservatione & curatione disputatio IV. Quam peculiari collegio, praeside Iohan. Iessenio a Iessen, Doctore & Professore. Ad Cal. Septembris adornat Daniel Sennert Vratislaviensis Sil.* (Wittenberg: Iohannes Dörffer typis Cratonianis, 1596). The printed disputation exists in a copy found in the Wroclaw University Library, shelfmark 388223. The 1599 disquisition has the following title: *Iohan. Iessenii a Iessen De sympathiae et antipathiae rerum naturalium causis disquisitio singularis. Quam in publico pro virili ad Cal. Iunij defendere conabitur M. Daniel Sennertus Vratislaviensis* (Wittenberg: Meißner, 1599).

34. Jessenius, *De sympathiae et antipathiae rerum naturalium causis*, B3r: "Cum enim elementorum vires excedunt effectus, ipsis nequaquam accepti ferendi. Nam cum omnis operatio e forma prodeat, neque quicquam ultra facultatem formaeque consequentiam possit agere, existimandum elementorum virtutes gradum suarum formarum non excedere, sed humida, quatenus talia, solum humiditatem, frigida frigiditatem imprimere: non item ferrum trahere, aut puppim detinere, quae Herculeus lapis, & Echeneis praestare assolent, quas vires simpliciter mistis his coelitus immissis arbitrandum. Idcirco lis & amicitia, quae ad elementorum Qualitates reduci nequit, superiorum efficaciae adscribenda." See Christoph Lüthy and Willam R. Newman, "Daniel Sennert's Earliest Writings (1599–1600) and their Debt to Giordano Bruno," *Bruniana and Campanelliana* 6(2000/2), pp. 261–279. For Sennert's early view on the subject of occult qualities, see his *Epitome*, 1599–1600, Disputation 11, theses 20 and 26.

35. Hiroshi Hirai, "Le concept de semence dans les théories de la matière à la Renaissance: De Marsile Ficin à Pierre Gassendi" (Ph.D. diss., Université Lille III, 1999), pp. 62–77; Copenhaver, "Astrology and Magic," pp. 286–287; Thorndike, *History of Magic*, vol. 5, pp. 557–560.

morbus Gallicus in 1530, in which he considered the causes of the disease that still goes by that name; in his *De contagione* of 1546, he wrote extensively on the theory that disease is transmitted by tiny "seeds" (*seminaria*) as opposed to being uniquely the product of humoral imbalance. Fracastoro's seeds were themselves mixed bodies, bearing both the "material" qualities of the elements and certain "spiritual" qualities. Although these "spiritual" qualities were not identical to the occult qualities of scholasticism—and indeed, Fracastoro was an avowed enemy of explanations involving occult qualities—the fact remains that Fracastoro, like Sennert, employed corpuscles that were themselves mixts as vectors of nonstructural properties.[36] Sennert received further knowledge of disease seeds or *semina* from Petrus Severinus, whom we encountered in a previous chapter as an interpreter of Paracelsus: the iconoclastic Swiss chymist had himself held an influential seed-based theory of disease.[37]

Hence Sennert was the heir of a well-established tradition that took the emphasis off of the elemental qualities embodied in the four humors of traditional scholastic medicine and advocated an ontological theory of disease. In the case of Fracastoro, Severinus, and others, the disease entities were even carried by vectors called "seeds"—highly reminiscent of the *semina* or atoms of the ancient follower of Epicurus, Lucretius. If we turn to Sennert's early medical disputations, the seriousness with which he took these ideas at once becomes apparent. A 1604 disputation on the method of curing (*De methodo medendi*) composed by Sennert explicitly equates Galen's properties of the whole substance and scholastic occult qualities. Sennert explains the ability of some medicines to purge on the assumption that the medicine attracts an offending humor due

---

In addition to these authors, there is a fairly extensive literature on Fernel, much of which is cited in Nancy Siraisi, "Giovanni Argenterio and Sixteenth-Century Medical Innovation: Between Princely Patronage and Academic Controversy," *Osiris*, 2d ser. 6 (1990): 161–180; see p. 161, n. 1. See also Linda Deer Richardson, "The Generation of Disease: Occult Causes and Diseases of the Total Substance," in *The Medical Renaissance of the Sixteenth Century* (Cambridge: Cambridge University Press, 1985), pp. 175–194.

36. See Hirai, "Le concept de semence," pp. 54–59. Hirai links the *qualitates spirituales* of Fracastoro with the *spiritus vitae* of Marsilio Ficino. See also Thorndike, *History of Magic*, vol. 5, pp. 488–497. For Fracastoro, see Vivian Nutton, "The Seeds of Disease: An Explanation of Contagion and Infection from the Greeks to the Renaissance," *Medical History* 27 (1983): 1–34; and Nutton, "The Reception of Fracastoro's Theory of Contagion: The Seed That Fell Among Thorns?" *Osiris*, 2d ser. 6 (1990): 196–234.

37. For Sennert's awareness of Severinus and his importance, see Jole Shackelford, *A Philosophical Path for Paracelsian Medicine: The Ideas, Intellectual Context, and Influence of Petrus Severinus: 1540–1602* (Copenhagen: Museum Tusculanum Press, 2004), pp. 310–314.

to a familiarity of substance between the humor and the drug. But form only acts by means of qualities, and it is the occult qualities emanating from the form that perform this admirable feat. Following Scaliger, Sennert adds that it is impudence to deduce all effects from the manifest elemental qualities hot, cold, wet, and dry—"for if this attraction were caused by the primary qualities, all things whatsoever would be endowed with such [an attractive] quality," since the four elemental qualities are present in everything that is made from the elements.[38]

The same affirmation of occult qualities appears in Sennert's disputations on the differences of diseases (*De differentiis morborum*) and on malignant fevers (*De febrium malignarum natura & causis disputatio*), dated 1605 and 1607, respectively.[39] The latter disputation is particularly interesting for its assertion that occult qualities, although admittedly a principle of nescience, are to be preferred over the fallacious attempt to reduce all phenomena to the four elemental qualities—"Others, lest they seem not to know something, attempt to deduce all things from the manifest qualities," and thus argue that pestilence arises from simple putrefaction alone. But they cannot account for the great variation in the virulence of pestilential fevers by granting them a single, universal cause, and so they resort to ad hoc explanations.[40] This is a theme that would occupy Sennert at much greater length in his disputation on pestilence, published in the same year (*De pestilentia disputatio*, 1607).

Early in the *De pestilentia disputatio* of 1607 Sennert reiterates the claim that pestilence cannot arise from the action of the four primary qualities alone. Although they do produce damage in many instances, as in consumption (*phthisis*), for example, when the patient's humors gradually dry up, they cannot account for the rapidity with which the plague acts. Pestilence works more in the fashion of a poison—killing rapidly and

38. *De methodo medendi disputatio vi. de purgatione: in qua, cum deo, sub praesidio Danielis Sennerti, medic. doct. et profess. p. respondentis partes sustinebit Conradus Schattenbergius, Flenspurgensis Holsatus. ad diem 31. Martij* (Wittenberg: Crato, 1604), theses 38–39.

39. *De differentiis morborum disputatio prima. Cujus theses, cum deo, sub praesidio Danielis Sennerti, phil. et medic. doctoris et profess. p. defendendas suscepit Martinus Boecherus Austriacus. Ad diem 19. Ianuarij* (Wittenberg: Johann Schmidt, 1605), theses 27, 30, 40–42; and De febrium malignarum natura & causis disputatio; quam, cum deo sub praesidio Danielis Sennerti d. et med. profess. pub. publico examini subjicit M. Michael Döring, Vratislaviensis, in auditorio medicorum, Ad diem 13. Februarij (Wittenberg: Martin Henckelius, 1607), theses 16–18.

40. Sennert, *De febrium malignarum*, thesis 17: "Alii, ne quid ignorare videantur, ex qualitatibus manifestis cuncta deducere annituntur, solique illa putredini accepta ferenda esse tradunt. Cum autem & hi videant, simplicem putredinem non sufficere, quippe ex qua rationem sufficientem redituri non sint, cur omnes febres putridae pestilentes haud sint atque malignae: nescio, quos non putredinis labyrinthos inexplicabiles comminiscantur."

experiencing no alleviation from traditional medicines. And if plague were caused by mere putrefaction, then why is it that other animals, which would also be subject to putrefaction, do not die when a plague epidemic is raging? For this and other reasons, Sennert acknowledges that pestilence must originate from an occult contagious venom, whose action is specific to the heart of man. But in response to the question that someone might pose—what is the specific nature of this poison, and how does it differ from others—Sennert replies again with an admission of nescience: "We respond with Scaliger, in *exercitatio* 218, section 8— 'these things are such as escape temperate minds and mock the curious,' and in *exercitatio* 307, section 29—'it is the lot of human wisdom to desire with a tranquil mind not to know (*nescire velle*) certain things.' "[41] The precise way in which the plague operates escapes our senses and hence our knowledge.

Nonetheless, the presence of contagion in a plague epidemic is obvious to everyone, and hence a vector must be present. At this point, Sennert adopts the theory of Fracastoro (and others) that plague travels by means of "seeds" (*seminia, seu seminaria*). These seeds carry the pestilence, which operates by means of its whole substance. Fernel is very likely right, Sennert says, in arguing that the occult qualities transmitted by such seeds stem from the celestial bodies. This explains why pestilence can occupy almost the whole world at once, as opposed to a disease brought on by local atmospheric putrefaction and subterranean exhalations. Although Sennert demurs from affirming the technical details of astrology, it is clear that he believed (at least in 1607) that pestilence derived from poisonous occult qualities carried on minute material seeds. The latter could be inhaled or, when the pores were open and relaxed, they could enter directly into the skin, whereon they would pass to the heart and attack it.[42]

The relationship between Sennert's developing interest in corpuscles and his reading of authors like Fernel and Fracastoro is straightforward. Although Sennert had not committed himself to atomism by 1607, he was deeply curious about occult qualities, and saw them as a means of escaping the jejune reduction of phenomena to the four elementary qualities. Fracastoro, in particular, had a well-developed theory

---

41. Sennert, *De pestilentia disputatio* (1607), thesis 21: "Si quis autem porro quaerat, quae veneni pestilentis sit specifica natura, & quomodo ab aliis venenis differat: ei cum Scaligero exerc. 218 sect. 8. respondemus; Haec talia esse, quae latent animos temperatos, illudunt curosis; & exerc. 307. Sect. 29. Humanae sapientiae partem esse; quaedam aequo animo nescire velle."

42. Sennert, *De pestilentia disputatio* (1607), theses 25, 26, 34, 44, 45, 54, and 58.

involving corpuscles as vectors of powerful active properties—why not then link these agencies to the occult qualities described by Fernel and a host of other Renaissance authors? Indeed, Sennert would pursue this tactic even further in the 1619 *De chymicorum*, where Fracastoro became a witness of the fact that there are subtle, particulate effluvia that escape our senses, and indirectly, that there are occult qualities.[43] The medical theory of "seeds" endowed with occult qualities clearly provided the young Sennert with a model for a corpuscular theory also incorporating occult qualities to account for the sympathies and antipathies between different substances. The great and seemingly instantaneous changes wrought by spirit of vitriol (sulfuric acid) when poured on oil of tartar (potassium carbonate), or by spirit of niter (nitric acid) when mixed with the oil of mercury and antimony, were derived from the combining and separating action of occult qualities that emanated from substantial forms locked within their atoms.[44] Although the medical tradition did not supply Sennert with the structural explanations of phenomena in terms of atoms or corpuscles undergoing *synkrisis*, *diakrisis*, and immutation that came from his Democritean and Geberian reading of chymistry, the scholastic physicians did provide the immaterial agencies by which the substantial forms of different things exercised their hidden activity. The occult qualities and the equally occult forms from which they flowed served as extrinsic limits of the observable phenomena, just as the procedures of the laboratory provided access to the intrinsic limits supplied by analysis.

It is obvious that the occult qualities and substantial forms populating Sennert's world acted not only as an acknowledgement of observational limits, but as a source of efficient causality.[45] The atoms involved in chymical operations derived the sympathy and antipathy leading to *diakrisis* and *synkrisis* from the occult qualities provided by their substantial forms. But this raises a serious problem. Where should one draw the line between perceptible qualities caused by structural factors and those flowing directly from the substantial form? We know that Sennert by 1619 (if not earlier) believed forms to be immutable, and yet in some instances, such as the transmutation of base metals in chrysopoeia, he explicitly says that forms can be removed and replaced.[46] How are we to interpret

43. Sennert, *De chymicorum* (1619), p. 252.
44. Sennert, *De chymicorum* (1619), p. 355.
45. See Michael, "Sennert's Sea Change," pp. 357–362, on this point.
46. Sennert, *De chymicorum* (1619), p. 24.

this claim? Did Sennert believe that a given corpuscle of silver could simply experience a corruption of its substantial form and the instantaneous acquisition of the form of gold *tout court*, or did he think that this substantial change required a simultaneous rearrangement or other structural alteration of the corpuscles themselves? In fact he is not often clear about this point, and one must suspect that he had not made up his mind, at least by 1619.

Indeed, in the very text where he gives a Democritean interpretation to the *synkrisis* and *diakrisis* of the chymists, Sennert also develops a sophisticated explanation of phenomenal change in hylomorphic terms—his theory of subordinate forms. As we have already had occasion to note, some scholastic authors postulated that a *forma mixti* could coexist with the forms of the elements in a mixture. Out of this tradition, especially from his ongoing dialogue with Zabarella and with the aid of speculations drawn from chymical sources, Sennert seems to have developed his theory of subordinate forms, whereby one form could persist while acting as the "matter" in which another form could inhere.[47] The subject arises in the *De chymicorum* in Sennert's discussion of the commonly occurring "degeneration" of wheat into the weed darnel—this was an old example of the transmutation of species that often appeared in alchemical texts and elsewhere. Sennert says that there are two possible explanations for this type of phenomenon—"either one form can go forth into the theater of nature under diverse bodily shape, or to be sure diverse forms can exist in one seed, but subordinated, so that one is the princess and lady, and the rest, as it were, are servants."[48] We have already met with the first of these two possibilities, when Sennert was describing the sensible

47. Michael, "Sennert's Sea Change," pp. 336–348. Michael is unaware of Sennert's clear debt to the chymists on this point, however. The immediate inspiration for Sennert's treatment of species degeneration appears to have been Severinus's doctrine of "transplantation" of species, according to which the *semina* of one species become subordinate in a particular body and those of another species become dominant. Severinus's terminology here is at times precisely that of the subordination of forms theory, as in the following example: "Sic in semine Tritici Forma Lolii delitescit, sed ministra, aequivoca, accidens." This is an interesting example of the way in which Severinus's *semina* theory could easily be reconfigured to serve as the basis of atomistic speculations. See Petrus Severinus, *Idea medicinae* (Basil: Henricpetrus, 1571), pp. 139–146; the quoted passage is found on p. 141. See also Shackelford, *A Philosophical Path*, pp. 183–185, for discussion of transplantation in Severinus.

48. Sennert, *De chymicorum* (1619), p. 336: "Cum multis, quae in natura fiunt, probabile reddatur, unam formam sub diverso schemate corporis in naturae theatrum progredi posse: Aut certe in uno semine posse esse plures formas, sed subordinatas, ita tamen, ut una sit princeps & Domina, reliquae quasi ministrae." Sennert proposes these two competing explanations again on p. 347.

alterations of mercury or of metals dissolved in acids as an alteration of external form while the internal form remained the same. This type of alteration, where no formal replacement occurred, committed Sennert to a structural explanation. The same unchanging form, locked in its corpuscular prison, could act as an efficient cause producing different states of aggregation with strikingly divergent qualities—Sennert's so-called "external forms." Since the "internal" or substantial form itself remained unchanged during these phenomenal mutations, the apparent changes in quality had to be due to alterations in the structural properties of the atoms (*diakrisis, synkrisis,* or immutation).

But what of the case where multiple forms are subordinated to one another, so that one substantial form rules at a particular time but may be supplanted by its erstwhile minion upon undergoing corruption? Sennert gives the example of plants putrefying into caterpillars, for in such a case one substance, the plant, is clearly replaced by another, namely, the in-sect. Although the conventional literature on this subject viewed such metamorphoses as cases of purely spontaneous generation due to the action of celestial influences on matter, Sennert argued that their regu-larity required an internal, species-determining agency. In fact, "spon-taneous" generation involved the actualization of a hidden subordinate form, for "certain *semina* are, as it were, ambiguous, and contain within themselves multiple forms."[49] In order to convince himself that such generations really could occur without parents or the putative celestial influences, Sennert performed an experiment, hatching maggots from red cabbage by means of artificial heating in midwinter, when there were neither adult flies available to lay their eggs nor the warming influx of the sun.[50] Clearly the maggots must have come directly from the plant, which contained "equivocal" *semina* whose subordinate forms were ac-tualized as the plant decayed.

Hence the fact that particular types of matter usually gave birth to specific types of life—as when a mulberry tree produced silkworms—supported Sennert's theory of subordinate forms. Spontaneous

49. Sennert, *De chymicorum* (1619), p. 347: "Deinde non plane a ratione & experientia alienum videtur, quod semina quaedam sint quasi ambigua & formas plures in se includant: ita tamen, ut una dominari debeat, altera servire; nisi occasione oblata, aliud fiat."

50. Sennert, *De chymicorum* (1619), pp. 337–338. I have put the word "spontaneous" in quotation marks to indicate Sennert's own unease with the term. Although he believed that lower animals and plants could be generated without parturition, he argued against the classical view that life could be generated spontaneously from the elements or other purely inanimate matter. Instead, some sort of "seed" (semen) had to be present in the matter from which the being came to be, but this "seed" could be invisibly small.

generation was not a chaotic and unguided development of life from putrefying matter. Rather, it was the unfolding of preexistent subordinate forms, once the substantial form that dominated them had been corrupted or otherwise lost its ability to rule. Now this fit very nicely with another of Sennert's beliefs, one that he had already tentatively proposed in the 1599–1600 *Epitome* but which he espoused fully as his own in the *De chymicorum* of 1619. This was the idea that form was not "educed" from matter on an ongoing basis but that God had made the species of all things in the original Creation and endowed all beings with forms that allowed them to multiply themselves.[51] These forms were durable and capable of multiplying themselves in matter; they were passed on from generation to generation rather than being transmitted to matter by the celestial bodies whenever a new mixture occurs, as Fernel thought.[52] In higher animals they exist in the semen, in lower animals within eggs, and in the beings that come to be by seemingly spontaneous generation, they are present as "seeds," "seminal principles," or "atoms." The appearance of worms in decaying flesh is due to the presence of such seminal principles in the flesh, where the latter were present all along but forced into subordination by the dominant form of the living animal—its soul—so long as it lived. It is the perseverance of such forms, whether they be substantial or subordinate, that accounts for the stability and longevity of God's creation.

Clearly Sennert's subordinate forms provide him with a means of avoiding the commonplace charge against ancient atomism that it could not account for the regularity of the world, since the atoms of the Greeks were unguided by any principle of design. By seeing the successive mutations of plants into animals as the unfolding of hidden, subordinate forms, one could retain the principle that form itself is immutable while also preserving the regularity of nature. As a guarantor of the world's regularity, form was "the instrument and hand of the wisest God," which He Himself must have imposed on matter at the beginning of time. Sennert has a great deal to say about the divine character of forms and of their role in the perpetuation of species.[53] My goal is not to pursue that

---

51. Sennert, *Epitome* (1599–1600), disputation 3, theses 41–42. Sennert, *De chymicorum*, 1619, pp. 330–331, 344, and passim. The idea seems patently related to Saint Augustine's concept of *rationes seminales*, but Sennert does not explicitly derive it from Augustine.

52. Sennert, *De chymicorum* (1619), pp. 191–193.

53. Sennert, *De chymicorum* (1619): p. 353: "Formae enim sunt principium divinum & immutabile, quod determinat omnes actiones & passiones rei naturalis; & sunt quasi instrumentum ac manus sapientissimi Creatoris ac Opificis Dei."

discussion deeply here, since it would entail a sustained consideration of Sennert's later work and a thorough analysis of his views on theology. It is necessary, rather, to emphasize two things. First, Sennert's exalted view of form as the direct gift of a Creator God, having been passed on since the beginning of time and remaining responsible for the activity of matter, clearly committed him *ab initio* to a sort of theistic hylomorphism. His was a "religion of form." Second, substantial form was immutable in Sennert's system. If, in his youth, he believed with Zabarella that the forms of the elements could be weakened or remitted, this was only because he also accepted the Averroistic notion at that time that the elemental forms were somehow intermediate between substance and accident. Had they obtained the status of full-blown substances, their intension or remission would have been inadmissible even to the young Wittenberg graduate. This impassibility was shared by the substantial forms of Sennert's maturity, of course, but even if they could not be intended and remitted, they could be imposed and removed. As Sennert pointed out repeatedly in the 1619 *De chymicorum*, then, there were two ways in which one could conceive of most phenomenal change. Either the substantial form remained constant and the material structure in which it inhered underwent structural alteration, or the dominating form itself was removed and replaced by a subordinate form. A third possibility also emerged in the context of metallic transmutation, where Sennert argued that one indwelling substantial form was removed and replaced by an extrinsic one.

But how could one determine which explanation of qualities—the structural or the hylomorphic—was actually correct in a given case? Interpretive difficulties arise from the fact that they are not necessarily mutually exclusive. Even in the case of purely structural change, brought about in instances of *synkrisis*, *diakrisis*, and immutation, the substantial form played a part by exercising attraction or repulsion through its occult qualities. We saw this clearly in Sennert's experimental reductions to the pristine state. In the two cases of formal replacement, on the other hand, either by a subordinate or by an extrinsic form, it would also be possible for Sennert to have argued that the succession of forms could only occur with a concomitant alteration of material structure. But he does not argue unequivocally for this, even in his late *Hypomnemata physica*. As late as 1636, Sennert was still juxtaposing hylomorphic explanations with structural ones in a way that he did not seem to conceive of as problematic, and if anything, his theory of subordinate forms finds greater expression in the *Hypomnemata* than in the *De chymicorum*.[54]

Nonetheless, the *Hypomnemata* does throw some light on Sennert's attempt to distinguish the activity of successive subordinate forms from the action of a single internal form capable of inducing matter to take on different external appearances according to its structural alterations. He argues there that the metamorphosis of a caterpillar into a chrysalis and finally into a butterfly cannot result from the actualization of subordinate forms, since no essential mutation has taken place. The insect has no more changed its species than a snake acquires a new species when it sheds its skin.[55] Subordinate forms cannot serve as an *explanans* when we know the species of a thing to be fixed. In this case, a single substantial form must be unfolding the multiple possibilities of its essence by altering its matter in various ways. But what of the traditional instances where wheat degenerates into darnel, or turnips into radishes? In reply to this question, Sennert provides an additional way of determining the applicability of the subordinate-form theory. In effect, what he suggests is a sort of reduction to the pristine state, transferred from the region of minerals and metals to the realm of the animate world:

> Therefore this is more reasonable, that the Creator granted to these forms the power of fabricating various bodies for themselves, so that a certain form, when its matter is rightly disposed, effects [a body] in the way commonly [known]. Hence wheat very often generates wheat. But when the matter is less rightly disposed, lest nature be idle, it has the power of producing another body, as when the soul of the wheat has a less well disposed matter, it makes the body of darnel, or of Emmer wheat. And the thing occurs in virtually the same way as we said before in caterpillars and butterflies, where the same form fabricates diverse bodies at diverse times. The fact that the soul does not perish appears in fact from this—that after the Emmer wheat is born from the wheat, or black oats from white, upon finding a suitable earth and matter, they turn back into wheat or white oats.[56]

54. Sennert, *Hypomnemata* (1636), pp. 70–73, 394–402. It is true, however, that Sennert places greater contrast on the immutable character of the metamorphosing silkworm's substantial form as opposed to its changing "external form" here than he does in the *De chymicorum* of 1619. This probably reflects the fact that he has grown bolder in his atomism and more willing to commit to an openly structural interpretation of change even in the case of a living being.

55. Sennert, *Hypomnemata* (1636), p. 399: "At quis dicat hic formam internam mutari; nisi etiam, dum serpentes senectam exsuunt, novam generationem fieri dicere velit."

56. Sennert, *Hypomnemata* (1636), p. 450: "Ideoque hoc magis vero consentaneum est, Creatorem formis illis vim varia corpora sibi fabricandi concessisse, sicut quaedam, cum materia

Just as in the case of the reduction of silver from nitric acid, the unchanging character of the wheat's substantial form, here identical to its soul, reveals itself from the fact that Emmer wheat can return to wheat when the proper material conditions are met. The same logic permeates Sennert's thoughts in the metallic and the animate realm. If a substantial form is corrupted, and hence absent, it can no longer act on matter to produce its characteristic qualities.[57] Hence the fact that these characteristic qualities can be regained after their disappearance from the silver or the wheat means that no such corruption, and hence no essential mutation, has occurred, but only a change of accidents. In such a case, the unchanged substantial form has merely altered its material *emboîtement* in two different ways. But if a substantial form had been corrupted and removed, while a subordinate form had been actualized so that the latter dominated over the matter at hand and became its new substantial form, a new species would have had to be imposed. In this fashion the reduction to the pristine state provides a decisive test for determining the applicability or nonapplicability of the theory of subordinate forms.

Hence we find Sennert returning once more to the reduction to the pristine state as a means of determining the characteristics of form in general. The *Hypomnemata physica* takes a somewhat harder line on the unchanging character of substantial forms in living beings than did the *De chymicorum*. But again, Sennert appeals to the same reversible reactions that he had long before employed to prove the existence of atoms and to rebut the Thomistic and Averroist theories of mixture. There is every reason to see the reduction to the pristine state, especially as it appears in the case of metals dissolved in acids, as a crucial experiment

---

est recte disposita, vulgariter efficiat. Ita ex tritico frequentissime triticum nascitur. Cum vero materia minus recte est disposita, ne tamen natura sit otiosa, vim habet aliud corpus producendi, v.g. cum tritici anima habet materiam minus recte dispositam, facit corpus lolii, zeae. Et habet sese res eodem fere modo, sicut antea quoque dictum, ut in erucis & papilionibus, ubi eadem forma diversis temporibus diversa corpora fabricatur. Et animam non perire, patet vel ex eo, quod, postquam zea e tritico nata, avena nigra ex alba, iterum convenientem terram & materiam invenit, rursum in triticum, vel avenam albam abit."

57. One might be tempted to reply that a weakened substantial form that has been pushed into a submissive status by an erstwhile subordinate form could emerge later, if reinvigorated, to reclaim its species-imparting character, but Sennert does not seem to think this to be a serious possibility. In his view, the subordinate form only emerges when the substantial form has been obliterated. In his treatment of fungi growing on trees, for example, he explicitly states that the fungus only grows on the parts of the tree where the soul of the tree has departed. See Sennert, *Hypomnemata* (1636), p. 426: "Non autem fit ex ea fungus, quandiu est sub dominio animae arboris, vel fruticis, sed ubi ab ea deseritur, & vel tota planta, vel pars eius emoritur."

for Sennert. He makes this quite explicit in his ongoing treatment of spontaneous generation, as when he considers the equivocal production of lower animals and plants. Worms and weeds do not just arise directly from the elements but come into being from the action of atoms bearing substantial forms. These atoms can remain latent for long periods in other matter without informing it. The clearest example of this ability of atoms to hide, Sennert says, occurs in the case of gold dissolved in aqua regia or of silver dissolved in aqua fortis. As he puts it, "[T]hey retain their forms intact as reduction teaches, and yet they do not inform the acids (*aquas*); rather the form is in the acid (*aqua*) as in a place. The same is also true in the souls of living beings."[58] Hence the invisible *semina* of frogs or plants can lie latent in matter just as the atoms of a metal hide invisibly in an acid. A few pages later in the *Hypomnemata*, Sennert returns again to this theme, adding that it is not absurd to believe that spontaneous generation is really due to invisibly tiny seeds, since the metallic atoms released by dissolution in acid are so small that they can pass through filter paper and yet can still be reduced intact into the formerly perceptible metal.[59] Dissolution and reduction reveal not only the unchanging character of the form-bearing atoms but also their amazing minuteness. Once again, the reduction to the pristine state provides Sennert with experimental evidence for determining the nature of forms far beyond the phenomena of the metallurgical or mineralogical laboratory.

CONCLUSION

In the final analysis, Sennert's philosophy of nature must not be seen as a mere prelude to the mechanical philosophy, despite the clear presence of structural reductionism throughout his mature work. It would be a crude historical fallacy to think of Sennert's corpuscular theory simply as a halfway house in the supposedly inexorable progress from scholastic hylomorphism to the mechanical theories of the mid-seventeenth

---

58. Sennert, *Hypomnemata* (1636), p. 387: "Et, quod clarissimum est exemplum, est in aqua Regis aurum in minima solutum, & in aqua forti argentum in minima solutum; ita tamen, ut formas suas, ut ex reductione patet, integras retineant; nihilominus aquas illas non informant, sed auri & argenti forma est in aqua illa, ut in loco. Idem & in animabus viventium apparet."

59. Sennert, *Hypomnemata* (1636), p. 411: "Neque id absurdum est, sed formas in minimis, & sensu non comprehensibilibus atomis integras remanere posse, vel metalla docent, quae in minimas atomos ab aquis fortibus solvuntur, ut etiam per chartam colari possint, & nihilominus in minimis illis atomis suam essentiam integram retinent, ut ex reductione patet."

century. There is no reason to think that Sennert was motivated by the same desire that drove Descartes and Boyle to abolish substantial forms in favor of an inert, homogeneous matter having only a few primary qualities. To the contrary, Sennert saw no need to eliminate or even to question the theory that the phenomena of our world required the interaction of matter and form for their explication. In fact, it is likely that Sennert would have rejected the Cartesian theory of matter as excessively aprioristic, if he had been exposed to it. His own version of corpuscularism forbade elaborate speculation about the shape and arrangement of invisible particles, partly because of his focus on experience and partly as a result of Sennert's debt to the alchemical tradition of Geber, which likewise eschewed such theorizing.

At the same time, however, Sennert was an acolyte of what I have called the religion of form. Being the hands of God, substantial forms were not to be disposed of. They were the instruments by which nature, "the ordained power of God," performed its work.[60] God Himself created them at the beginning of time and put them in matter so that they could multiply themselves and maintain the species that He imposed on matter. Sennert himself did not view substantial forms as an unfortunate but necessary consequence of scholastic hylomorphism—rather, they were active witnesses of God's power and beneficence in the world. Once we acknowledge this fact, Sennert's experiments with the reduction to the pristine state acquire an added significance. It is true that he used these experiments in a decisive fashion to eviscerate the scholastic theories of mixture that relied on the corruption of old forms and imposition of new ones and as evidence for the reality of semipermanent corpuscles beneath the level of sense. Yet at the same time, Sennert's explanation of chymical reduction locked the forms safely within their material vehicles, the atoms, and allowed them to persist in the face of the technological assault stemming from those striking agents of qualitative change, the mineral acids. The reversible reactions wrought by powerful acids and bases gave no quarter to the tradition inaugurated by Thomas Aquinas with its claim that the ingredients of a mixture could not be recaptured. From Sennert's perspective, then, his atomism had saved the substantial form from the unworkable theorizing of an excessively metaphysical type of Aristotelianism that mired itself in speculation to the detriment of the evidence.

---

60. Sennert already employs the expression "ordinaria Dei potestas" for nature in the 1599–1600 *Epitome*, in Disputatio IV, Thesis II. He attributes it there to Scaliger. The expression appears later throughout his works, as at *Hypomenmata* (1636), p. 2.

But Sennert's confidence, for all his success in building a corpuscular theory on the ashes of late Renaissance scholastic concepts of mixture, contained an unwitting irony. His reliance on the *diakrisis, synkrisis,* and immutation of minute atoms to explain a host of visible changes shifted the burden of explanation away from immaterial form to the structure of the matter in which it inhered. All qualities flow from form, as Sennert himself would say, and yet the opaque and shiny tint of silver disappeared when the metal dissolved into a clear solution in nitric acid, as he observed no less astutely. The elimination of the silvery color of the metallic mass resulted from the mere fact that the silver had been divided into particles that were too small and dispersed to see, not from the loss of an old form and the imposition of a new one. The same demotion of the substantial form as an explanatory agent can be seen in most of Sennert's explanations involving *diakrisis* and *synkrisis.* Hence in saving the substantial form, Sennert at the same time rendered it largely otiose as a direct source for the qualities that were supposed to flow from it. Substantial forms acquired a new emphasis as efficient causes of corpuscular local motion in his system, but in the process they lost a corresponding portion of their explanatory power as an immediate source of qualitative change, and hence of quality itself. It was the birthright of a later generation, no longer wedded to the scholastic Aristotelianism of the early modern German university, to see this chink in the armor with which Sennert had clothed his substantial forms. He had provided the very means of making forms unnecessary, and it was merely by drawing out the consequences of his own reasoning that Sennert's system could be subverted. This task would become a veritable *ideé fixe* in the hands of Sennert's most influential heir and critic, Robert Boyle.

THREE

Robert Boyle's Matter Theory

6

Boyle, Sennert, and
the Mechanical Philosophy

Traditional accounts of Robert Boyle's matter theory, such as Marie Boas Hall's 1952 *Establishment of the Mechanical Philosophy*, explicitly view Boyle's mechanical philosophy as an importation from physics, which he grafted onto a radically rewritten chemistry. As Boas Hall puts it, Boyle's "new chemistry" was "a chemistry in which was incorporated a physicist's view of matter."[1] The physicist's matter theory refers, of course, to the very corpuscularian philosophy to which Boyle devoted his life's work, the explanation of phenomena in terms of matter and motion at the microlevel. According to Boas Hall, this physicist's theory was radically opposed to the chymical theory that predated Boyle and that he sometimes criticized—particularly the theory of three principles, mercury, sulfur, and salt, invented by Paracelsus in the early sixteenth century. The Paracelsian concept of the *tria prima* was, to paraphrase Boas Hall, a theory of forms and qualities, an animistic rewriting of Aristotle in the language of alchemy.[2] A brief glance at Steven Shapin's 1996 *The Scientific Revolution* will show that the approach of Boas Hall is alive and well. In his treatment of the mechanical philosophy as a whole,

---

1. Marie Boas Hall, *Robert Boyle and Seventeenth Century Chemistry* (Cambridge: Cambridge University Press, 1958), p. 75.
2. Boas Hall, *Robert Boyle*, pp. 82–84.

Shapin places Boyle in the company of Democritus, Epicurus, Galileo, Bacon, Mersenne, Descartes, and Gassendi—precisely the representatives whom Boas Hall had in mind: the chymical writers are conspicuous only by their absence.[3] This is no surprise, since Shapin's bibliographic essay at the back of his book gives Boas Hall pride of place as the basic work on the mechanical philosophy.[4] But a more careful look at the development of Boyle's mechanical philosophy will show, pace Boas Hall and Shapin, that the chymical literature written by his predecessors played an essential role from the beginning and continued to be highly significant for Boyle's matter theory even in his maturity. Not only did traditional chymistry affect Boyle's thoughts on chrysopoeia and the presence or absence of seminal principles in matter (the offspring of earlier chymists' *semina*); it also deeply conditioned the mechanical philosophy that is perhaps Boyle's most enduring claim to fame.[5] Although Boyle's mechanical philosophy certainly owed a heavy debt to other thinkers in the "new science," particularly Gassendi and Descartes, the fact is that the main experimental support for Boyle's matter theory flowed in a direct line from the scholastic alchemical tradition articulated and refined by Daniel Sennert. Nor was this a case of Boyle's extracting brute matters of fact from chymical empirics in order to support his own explanatory theory: important components of the mechanical philosophy were already being used by chymists in the experimental contexts where Boyle found them.[6]

The major early source for Boyle's experimental support of his corpuscular philosophy—and a highly important one throughout his career—was Sennert's experimental reduction to the pristine state. Like Sennert, Boyle would use this phenomenon as direct evidence for the persistence of semipermanent corpuscles during mixture. Still, if we

3. Shapin, *Scientific Revolution*, pp. 49–52.

4. Shapin, *Scientific Revolution*, p. 174.

5. For Boyle's work in chrysopoeia, see Lawrence M. Principe, *The Aspiring Adept: Robert Boyle and His Alchemical Quest* (Princeton: Princeton University Press, 1998; for Boyle on seminal principles, see Antonio Clericuzio, "A Redefinition of Boyle's Chemistry and Corpuscular Philosophy," *Annals of Science* 47 (1990): 561–589, Jole Shackelford, "Seeds with a Mechanical Purpose: Severinus' Semina and Seventeenth-Century Matter Theory," in *Reading the Book of Nature: The Other Side of the Scientific Revolution*, ed. Allen G. Debus and Michael T. Walton (Kirksville: Sixteenth-Century Journal Publishers, 1998), pp. 15–44, and Peter Anstey, "Boyle on Seminal Principles," *Studies in History and Philosophy of Biological and Biomedical Sciences* 33 (2002): 597–630.

6. See Newman and Principe, *Alchemy Tried in the Fire*, pp. 15–34, 208–236, and 268–272, where we argue that Boyle systematically erased his debts to earlier chymistry despite the crucial theoretical and practical components that he borrowed therefrom.

compare Sennert's corpuscular views with those of Boyle, we see not only striking parallels but also remarkable divergence. As Thomas Kuhn already pointed out in his 1952 article, "Robert Boyle and Structural Chemistry," Sennert believed in the reality of immaterial substantial forms that were immutable and that had lost their ability to explain many simple chemical operations without the aid of additional suppositions.[7] Since an increasing number of chemical reactions were revealing themselves to be reversible and the prevailing scholastic theories of mixture could not adequately account for the reacquisition of ingredients from a mixt, Sennert's system can therefore be seen partly as an attempt to save the reality of substantial forms in the face of new analytical technologies that threatened their credibility. Now obviously this cannot be said for Robert Boyle, whose mechanical philosophy had precisely the goal of eliminating substantial forms from the explanatory armory of natural science. This leaves us, then, with a remarkable situation. Boyle borrowed Sennert's decisive experiment for the existence of atoms, the reduction to the pristine state, and used it as the centerpiece of his own attempt to debunk the very theoretical entities that Sennert was trying to defend.

Thanks to the recent work of Michael Hunter, Lawrence Principe, and others, we now have a good sense of the young Robert Boyle's transformation from a writer of romances and moral epistles to a preeminent member of the scientific community in seventeenth-century England.[8] Boyle's earliest serious scientific interest was undoubtedly in the realm of chymistry, in which he received his most important tutoring during the early 1650s from the remarkable New England émigré George Starkey.[9] At the same time, however, Boyle was already developing a generalized interest in the legacy of Francis Bacon, largely imbibed from the loose circle of intelligencers, projectors, technicians, and amateur scientists centered on the German expatriate Samuel Hartlib. Boyle's early Hartlibian associations, chronicled by Charles Webster, surely contributed to the full-blown Baconian empiricism that Rose-Mary Sargent

7. Kuhn, "Robert Boyle and Structural Chemistry," p. 24.
8. Michael Hunter, "How Boyle Became a Scientist," *History of Science* 33 (1995): 59–103; Lawrence M. Principe, "Style and Thought of the Early Boyle: Discovery of the 1648 Manuscript of *Seraphic Love*," *Isis* 85 (1994): 247–260; Principe, "Virtuous Romance and Romantic Virtuoso: The Shaping of Robert Boyle's Literary Style," *Journal of the History of Ideas* 56 (1995): 377–397; Principe, "Newly Discovered Boyle Documents in the Royal Society Archive," *Notes and Records of the Royal Society* 49 (1995): 57–70.
9. Newman, *Gehennical Fire*, pp. 54–91, Newman and Principe, *Alchemy Tried in the Fire*, passim.

has recently shown to characterize Boyle's mature natural philosophy.[10] At some point after his education in chymistry was well underway, Boyle composed a treatise on atomism (probably in the mid-1650s). A number of leaves from his manuscript treatise on the subject survive, despite the fact that Boyle later declared that this document was "to be rifled & burn'd."[11] Boyle's juvenile essay *Of the Atomicall Philosophy* contains his first unequivocal statement of adherence to a corpuscular theory of matter grounded on chymical experiment, which he would later make his major scientific goal.

In this early manifesto of atomism, Boyle borrowed heavily, and without acknowledgement, from the works of Daniel Sennert. Instead of citing the German academic, Boyle began the treatise by celebrating the recent resurgence of atomism, which he says is "so luckyly reviv'd & so skillfully celebrated in divers parts of Europe by the learned pens of Gassendus, Magnenus, Des Cartes & his disciples [and] our deservedly famous Countryman Sr Kenelme Digby."[12] This list, consisting exclusively of post-Sennertian natural philosophers and physicians, seems designed to advertise Boyle's adherence to the most up-to-date and novel form of "the new philosophy." In his later writings Boyle would portray Sennert as a learned physician, as a representative of the Paracelsian principles of mercury, sulfur, and salt, and as a scholastic, but he doggedly refused to admit Sennert into the camp of the new corpuscularian philosophy.[13] We are therefore confronted with a twofold problem—what precisely did Boyle acquire from Sennert's atomism, and how did he transform the Wittenberg scholar's ideas to make them conform to a nonhylomorphic corpuscular theory that was rapidly becoming "the mechanical philosophy"?

10. Charles Webster, *The Great Instauration* (New York: Holmes & Meier, 1976); Rose-Mary Sargent, *The Diffident Naturalist: Robert Boyle and the Philosophy of Experiment* (Chicago: University of Chicago Press, 1995), pp. 42–61. I use the term "Baconian empiricism" advisedly. As Sargent points out, Boyle's empiricism must be construed in such a way that it includes the generalized experience of legal cases and the "aids to the senses" provided by sustained experiment.

11. Boyle, *Of the Atomicall Philosophy,*" in Hunter and Davis, *Works*, vol. 13, xlii. Hunter and Davis are not entirely sure that this note is in Boyle's hand, but its presence, along with another to the effect that the treatise is "without fayle to be burn't," suggest strongly that it reflected his wish.

12. Boyle, *Atomicall Philosophy*, Hunter and Davis, *Works*, vol. 13, p. 227.

13. In the many places where Boyle actually does cite Sennert, he typically presents him as scholastic, chymist, and physician, not as a corpuscularian. In only one out of fifty-two references does Boyle even intimate that Sennert upheld a corpuscular theory of matter, and there Boyle misrepresents Sennert as an opponent of chrysopoeia. See Newman and Principe, *Alchemy Tried in the Fire*, pp. 20–21.

In order to address these questions, we will be forced to ask still others. Clearly, without some idea of what Boyle meant when he later used the term "mechanical philosophy," we will be at a loss as to how he transformed Sennert's ideas. We must therefore descend into the vexed issue of what the mechanical philosophy meant to Boyle. Our previous analysis of Sennert's work places us in a good position to do this, since it is quite obvious that Boyle meant to distance himself publicly from Sennert's hylomorphic ideas and used him as a conspicuous foil. Nonetheless, it is not my goal to provide a definition of the "mechanical philosophy" over the development of Boyle's work, and it is unlikely that such an approach would be successful, thanks to the nature of Boyle's own highly unsystematic thought, which does not lend itself to hard and fast definitions. A better way of approaching the issue is to determine what things Boyle found intriguing about the term "mechanical," and then try to see how these fit his theory of matter. This will be an exercise in teasing out associations rather than an attempt to arrive at a definition. Even if we cannot provide a positive definition, it may be possible to clarify Boyle's use of the term and to exclude some modern interpretations of his thought that are excessively nice in their attempt to distinguish "non-mechanical" components from the "mechanical" ones in his writings.

What we will find above all is that Boyle borrowed and adapted Sennert's use of the reduction to the pristine state and employed it in ways that Sennert himself had not dreamed of. There is, in fact, a considerable irony in this, since the German physician had viewed the reduction to the pristine state as the primary means of establishing the reality of semipermanent substances equipped with their own substantial forms. As we will see, Boyle ultimately used the same reversible reactions and others to demonstrate that the substantial form was an unnecessary assumption and hence an entity to be discarded. Nonetheless, he shared Sennert's belief that the reduction to the pristine state provided the surest evidence of material particles that could resist the assaults of powerful analytical agents and therefore supplied an empirical basis for the claim that matter was made up of corpuscles that did not, in the ordinary course of nature, lose their identity.

Our treatment of Boyle will therefore proceed in the following fashion. First, in order to confirm the fact that Boyle's corpuscular theory did owe a fundamental debt to Sennert, we will consider the textual evidence provided by the *Atomicall Philosophy* and a few passages from Boyle's *Sceptical Chymist*. Some of this will be familiar ground, since I demonstrated Boyle's debt to Sennert's experimental corpuscularism in 1996, but I will here add an analysis of the *Atomicall Philosophy's* arguments that will show precisely how Sennert's work fit into Boyle's experimental program for confirming the reality of corpuscles at the microlevel.[14] After this, we will pass to a consideration of the term "mechanical philosophy" in Boyle in order to assess the difference between his matter theory in the *Atomicall Philosophy* and his fully mature natural philosophy. Here we will rely on works that Boyle published mainly in the 1660s and early to mid-1670s, which can be seen as forming a group.[15] The issue of historiographical anachronism is deeply problematic in the case of a writer like Boyle, whose ideas developed in complex ways and over long periods of time. Works published after Boyle's *Mechanical Qualities* of 1675–76 will therefore be excluded from our consideration.[16] Finally, we will return to the issue of the reduction to the pristine state in Boyle's works, again restricting our analysis to the texts of the foresaid period. Having laid the groundwork by analyzing the *Atomicall Philosophy* and discussing the meaning of "the mechanical philosophy," we will be able to show how Boyle continued to use Sennert's demonstration of the existence of enduring corpuscles but in ways that were alien, and ultimately inimical, to Sennert's own project.

BOYLE'S EARLY USE OF SENNERT

As we saw above, Boyle opens his *Atomicall Philosophy* with an appeal to contemporary corpuscular theorists such as Descartes, Gassendi, and other notable figures in the "new philosophy." Then, like Sennert in his *De chymicorum* and *Hypomnemata physica*, Boyle argues that Aristotle misrepresented Democritus and other ancient representatives of atomism as point atomists. The ancient atomists, according to Boyle, intended their atoms to represent a lowest natural threshold of size. Although their atoms were extended, Boyle takes it as self-evident that nature could not

14. Newman, "Alchemical Sources," pp. 567–585.
15. See Hunter and Davis, *Works*, vol. 8, pp. xxix–xxx, where they point out that much of Boyle's *Experiments, Notes, &c. about the Mechanical Origine or Production of Divers Particular Qualities*, published in 1675–76, was probably already composed by the mid-1660s. They also note that Boyle's "History of Particular Qualities," published as part of his *Cosmical Qualities* in 1670, belongs to the same group of writings. For further comments on this compositional group, see their introduction to vol. 6, pp. xxxix–xl.
16. *Mechanical Qualities* is the short title given by Hunter and Davis to Boyle's *Experiments, Notes, &c. about the Mechanical Origine or Production of Divers Particular Qualities*. Hereafter, I will use the short titles given by Hunter and Davis for all of Boyle's works. See Hunter and Davis, *Works*, vol. 1, pp. xvi–xx for the complete list.

divide them beyond a certain atomic size. After Boyle's brief introductory remarks, he then jumps into a series of examples that are intended to demonstrate the probability of the corpuscular hypothesis. Here his debt to Sennert begins to emerge, as in the following example:

> In similar bodyes ⟨that are really such for wine milke &c. that seeme so are not⟩ their being constituted by Atomes is very probable, since it is so that their particles are very small & of the same nature with the whole they compose.[17]

A close inspection of this seemingly unassuming passage reveals interesting surprises. Boyle addresses himself first to the genuine "similar bodyes," among which his text includes the metals. These similar bodies are the homoeomerous substances of Aristotelian natural philosophy, even though Aristotle does not view them as consisting of atoms. As we discussed in our treatment of Sennert, Aristotle had argued in *De generatione et corruptione* (I 10 328a 11–12) that true mixture (*mixis*) comes about only when substances are combined throughout—"so any part of what is blended should be the same as the whole" (Forster's translation). As we know, Aristotle contrasts such true mixture with *synthesis* or juxtaposition, which consists merely of placing bodies side by side. Stock examples in the Aristotelian corpus of substances that have undergone true mixture include the metals, flesh, blood, wine, and milk.[18] In such mixed substances, every part will be alike, and the resulting mass will therefore be homogeneous. In short, to Sennert and the young Boyle, each atom making up the whole is itself a mixt, bearing the essential characteristics that brand the whole as a substance.

The first surprise lies in the fact that Boyle, the professed atomist, is willing in principle to accept that there are such homoeomerous substances marked by different essences. At this early phase of his career Boyle abandons the ontological simplicity of the Democritean atom, as Sennert had done before him, and treats individual atoms as if they were Aristotelian mixts. This move decisively distinguishes Boyle's early atomism both from the corpuscularism of Descartes with its undifferentiated first matter, and from the adherence to a "uniform catholic matter" of Boyle's maturity. Nonetheless, as we will soon see, it is the semipermanent, "atomic" nature of precisely such homoeomerous particles that

17. Boyle, *Atomicall Philosophy*, in Hunter and Davis, *Works*, vol. 13, p. 228. The editors of the new edition of Boyle's *Works* use angle-brackets to indicate Boyle's own authorial insertions.

18. See Aristotle *Meteorologica* IV 383a 21–22, 384b 30–34, 385a 8–10, 388a 13– 388b10, 389a 7–23, 390b 14–22, and passim.

will provide the mature Boyle with the primary redoubt in his defense of the mechanical philosophy.[19]

A second surprise arises from Boyle's causal assertion linking the homoeomerity of some substances with the fact that they are atomic. It is "very probable" that some substances are composed of atoms, "since it is so that their particles are very small & of the same nature with the whole they compose." The reader may justly wonder at the sense of this. Why should the uniform character of a substance support a belief in atomism? After all, Aristotle had, in his *De generatione et corruptione*, argued against atomism precisely because it violated the principle of uniform, perfect mixture. The answer here lies in Boyle's use of the term "Atomes," which he derives directly from Sennert. In this passage of the *Atomicall Philosophy*, the primary sense of "atom" does not refer to a body that is incapable of further physical division into smaller bodies. Boyle is not thinking here in the spatial, even geometrical, terms of classical atomism, even though he has just argued on behalf of Democritus that Aristotle was wrong in representing the Abderite as a point atomist. Instead, "atom" here refers primarily to a substance that is incapable of decomposition into more primitive components. Like Sennert, Boyle is reasoning within the framework of the "negative-empirical concept" by which a substance is viewed as elementary if it cannot be decomposed by the tools of the chymist. Within this framework, "elementary" and "atomic" can be seen as coextensive terms—both imply resistance to decomposition.[20] For this reason, then, the ability of some corpuscles to withstand the corrosive *menstrua* of the seventeenth-century chymist makes it probable that they are fundamentally indissoluble, and hence,

19. With regard to the mature Boyle, I mean "homoeomerous" in the sense that Geber used the term; on this see the treatment of Geber's *Summa perfectionis* in chapter 1. In Boyle's juvenile *Atomicall Philosophy*, to the contrary, he seems to have been willing to think of homoeomerous substances ("similar bodyes") as being perfect mixts.

20. As I have already pointed out, the notion of "atom" that Boyle inherits from Sennert is closely related to the chemical atomism of the late eighteenth and nineteenth centuries. Such atomism retained the concept of fixed chemical species and reduced microstructural speculation to the bare minimum. It is not the case, of course, that semipermanent material corpuscles are really required in order for something to be irreducible and intransmutable (hence "elementary"). One need only think of the spectral colors. Despite Isaac Newton's personal commitment to a corpuscular ontology, his claim for the elementary character of the spectral colors required only that he demonstrate them to be caused by rays of unequal refrangibility, not that they be corpuscular. The claim that matter is corpuscular in nature receives support in the work of Boyle and Sennert from a host of additional empirical considerations beyond mere resistance to dissolution, such as the ability to penetrate the fine pores of filter paper, the fact that sublimation often produces finely divided powders, and so forth.

atomic. The identity between the smallest attainable particles and the substance as a whole provides direct evidence for such atomism, since an obvious consequence of this uniformity is that no decomposition products are present.

Equally interesting is the fact that Boyle accepts only some of Aristotle's homoeomerous substances as being genuinely uniform. He explicitly rejects the homogeneity of wine and milk, for example. In fact, Boyle is probably deriving this correction of Aristotle from Sennert as well, who in the *Hypomnemata physica* asserted that milk and blood are actually heterogeneous at the microlevel:

> And even though milk may seem one body, its whey, butter, and cheese reveal [the existence] of diverse parts mixed *per minima*, when they are separated. So, too, even if the blood of animals appears to be one homogeneous body, not only are diverse parts found to exist in it which supply food to the various members of the body, but if it should be distilled, a volatile salt which was not in evidence before adheres in great quantity to the flask.[21]

Wine too consisted of heterogeneous particles according to Sennert: it was this that accounted for the separation of tartar from the wine by mere settling and the extraction of alcohol by means of distillation.[22] To Sennert, such substances as wine, blood, and milk are made of corpuscles that have, in Geberian fashion, been mixed *per minima* of smaller corpuscles and possess a certain amount of interparticular coherence. Hence he views the separation of milk into whey, butter, and cheese, the destructive distillation of blood, or the distillation of spirit of wine to have separated the preexisting particles of the fluid.

But the resemblance between Boyle and Sennert becomes even more pronounced when Boyle returns to the "similar" or genuinely homoeomerous substances, such as the metals. It is these that will give him experimental evidence of atoms, for they can only be broken into

21. Sennert, *Hypomnemata* (1636), pp. 113–114: "Lac etiam ipsum etsi unum corpus apparet: tamen diversas partes per minima mistas demonstrant, serum, butyrum, caseus, ubi separantur. Ita sanguis animalium etsi unum corpus homogeneum apparet: tamen non solum diversae in eo partes, quae diversis membris alimentum praebent, reperiuntur, verum etiam, si destilletur, sal volatilis, qui antea non apparebat, magna copia recipienti adhaerescit."

22. It is true, nonetheless, that Sennert says elsewhere in the *Hypomnemata physica* (p. 142) that wine is entitled to be called a "perfect mixt" as opposed to a mere juxtaposition. His point, however, is that the corpuscles making up the wine exhibit coherence and are not like Aristotle's wheat and barley shaken in a jar. Like Scaliger, Sennert was willing to continue employing the concept of the *mixtum perfectum*, but only in this restricted sense, which corresponds to Geber's "very strong composition" (*fortissima compositio*).

bits that are of the same nature as the whole, at least when applying the available agents of analysis. Unlike seemingly homogeneous materials like milk and wine, which have particles that are only loosely combined, the corpuscles of metals resist decomposition into simpler components, making them *a-tomos* or operationally indivisible. This linkage between indivisibility and permanence of identity would provide one of the enduring leitmotivs of Boyle's career-long attempt to justify the mechanical philosophy. His countless later attempts to induce and remove the perceptible qualities of materials without altering their essential nature rely, for the most part, on the use of substances that are, at least in the qualified sense that Geber used the term, homoeomerous. While the atoms of the *Atomicall Philosophy* are operationally indivisible and homoeomerous, however, Boyle adds that "their particles are very small," fulfilling a second obvious criterion of atomism. As we will see, he demonstrates this minuteness by employing Sennert's laboratory methods, just as he had borrowed the Wittenberg scholar's theoretical apparatus.

For his experimental evidence, Boyle thus descends to the particular example of metals dissolved in aqua fortis. In a passage of key importance, he reproduces Sennert's classic reduction to the pristine state, describing the dissolution of silver to produce a completely clear liquid, the passing of this solution through cap paper, its subsequent precipitation by salt of tartar, and the final reduction of the precipitate to regain the original silver. This proof of the minuteness and durability of the dissociated silver corpuscles is immediately followed in Boyle's *Atomicall Philosophy* by another metallic example. In the second example, Boyle fuses gold and silver together to make electrum, producing, as he says, "an union per minima that is Atomes." He then dissolves a portion of the alloy in aqua fortis. The aqua fortis, which can dissolve silver but not gold, allows the more precious metal to precipitate while the silver goes into solution. To Boyle this provides evidence that the apparently perfect mixture of the two metals was only an illusion: the alloy is really composed of distinct but cohering atoms that the aqua fortis accordingly separates. Let us quote the passage in its entirety:

> [T]hus sylver being dissolv'd in Aqua fortis & that Menstruum so well filter'd that the dissolv'd silver & it will both passe thorough Cap paper all the invisible particles of ye Metall which are so small that they hinder not the Diaphaneity of the Menstruum are yet each of them true silver as appeares by precipitating them to the bottome (by a little resolv'd salt of Tartar) in the forme of a subtle

powder which is easily reducible into the same numericall silver that was at first corroded & so in the mixture of Metalls there is an union per minima that is Atomes, as if gold & silver be duly melted together each part of the masse has an equall proportion of the respective Metalls, & any part of it being cast into Aqua fortis (which by reason of the virtue wee are now going to ascribe to it is by the French often call'd eau de depart) or water of separation, the Menstruum will corrode and imbibe the Atomes of [the] silver & let those of the gold fall in the forme of powder to the bottome, instances of this Nature might be easily multiplyed if I judg'd them requisite.[23]

The reader will at once recognize the origin of Boyle's two examples of reduction to the pristine state—they are taken almost verbatim from Sennert's *Hypomnemata physica*, where they appear, juxtaposed, in the same order in which Boyle gives them. All the elements of Boyle's demonstration are present in Sennert's work—the dissolution in aqua fortis, the clarity of the solution, the filtration, the precipitation, and finally the reduction of the precipitate to regain the original metal. And Sennert, like Boyle, intends all of this to prove the existence of unchanged metallic corpuscles. As the Wittenberg doctor says, upon reduction "the liberated atoms of the metals are united on account of their similarity and thus return to their original body."[24] At the simplest level, we see the dissolution of the gold-silver alloy being used by Sennert to demonstrate precisely the same result as in Boyle. As Sennert says, the gold and silver are mixed *per minimas atomos* to the degree that their identity is no longer perceptible. Yet the metallic atoms still retain their own essential characteristics, which can be demonstrated upon their parting by aqua fortis and subsequent reduction.

There is one highly significant difference between the two accounts, however. In the *Atomicall Philosophy*, Boyle has no desire to attack the scholastic theory of resolution to the four elements or the prime matter, as Sennert had done. What had provided Sennert with compelling evidence against the mixture theories of the Thomists and Averroists had become, in Boyle's hands, a bald affirmation of atomism. Having appropriated the groundwork, Boyle could afford to dispense with the superstructure. Nonetheless, in *The Sceptical Chymist* and *The Origine of Formes and Qualities*, Boyle would reintegrate a Sennertian attack on scholastic mixture theory into his discussion of the reduction to the pristine state,

23. Boyle, *Atomicall Philosophy*, in Hunter and Davis, *Works*, vol. 13, p. 228.
24. Sennert, *Hypomnemata* (1636), p. 111: Liberatae metallorum atomi ob similitudinem uniuntur, & ita in pristinum corpus abeunt.

where this would become part of a larger strategy to dispense altogether with substantial forms.[25]

That Boyle is indeed employing Sennert in the *Atomicall Philosophy* and not some other source is confirmed not only by the generally close parallelism between the two documents but also by the fact that both Boyle and Sennert use the two independent examples of an unalloyed metal and then a gold-silver alloy in the same order and without any break in their respective texts. If Boyle were cobbling his metallurgical demonstrations together from different texts, we would hardly expect the same examples to fall spontaneously into the same uninterrupted order. It is far more likely that he has extracted both examples from the same source, namely, Sennert. The evidence that Boyle was using Sennert here is further bolstered by the fact that we know from other sources that he was reading the German academic in the mid-1650s.[26] Boyle's debt is also revealed by terminological considerations. Although the verbal similarity between two authors writing in different languages must necessarily be limited, one can see Boyle borrowing technical expressions from Sennert's corpuscular repertoire, such as the term "union per minima" for an intimate mixture of particles (which, strictly speaking, is still not Aristotelian *mixis*), in accordance with the tradition of Geber. Other examples of such unacknowledged linguistic borrowing from Sennert can be found in Boyle's oeuvre as a whole, but it is more important that we now turn from the evidence of Boyle's borrowing and consider the use that he makes of Sennert's reductions to the pristine state.[27]

The significance of Boyle's recapitulation of Sennert here can be appreciated fully only if one considers its placement in the *Atomicall Philosophy*. Following Boyle's rejection of Aristotle's arguments against atomism and the latter's specious notion that milk and wine are perfect mixts, the

---

25. Boyle, *Sceptical Chymist*, in Hunter and Davis, *Works*, vol. 2, pp. 254–276, especially p. 269; *Forms and Qualities*, in Hunter and Davis, *Works*, vol. 5, p. 396.

26. See the discussion of Boyle's many references to Sennert in Newman and Principe, *Alchemy Tried in the Fire*, pp. 18–27. Boyle was already well acquainted with Sennert's work early in his own career, since he cites the German physician in the juvenile part 1 of *Some Considerations Touching the Usefulnesse of Experimentall Naturall Philosophy* and in his unpublished *Essay Of the holy Scriptures*, composed between 1649 and 1653. See Boyle, *Usefulness of Natural Philosophy*, in Hunter and Davis, *Works*, vol. 3, p. 254. For the date of Boyle's *Essay Of the holy Scriptures*, see Michael Hunter, "How Boyle Became a Scientist," *History of Science* 33 (1995): 59–103; especially p. 67; for Boyle's reference to Sennert therein, see p. 77.

27. See Newman, "Alchemical Sources," pp. 580–583, for further evidence of Boyle's debt to Sennert.

reduction to the pristine state provides Boyle's first experimental evidence that genuine homogeneous substances, such as metals, are "constituted by Atomes." Indeed, if we consider the *Atomicall Philosophy* as a whole, it is striking that Sennert's experiment provides the *only* direct evidence for the existence of atoms to be found in the text. The rest of the *Atomicall Philosophy* is concerned with determining the properties of atoms rather than proving their existence.

Following his recapitulation of Sennert's experiment, Boyle proceeds in section 2 of the *Atomicall Philosophy* to argue that "the almost infinite subtilty or smallnesse that they ascribe to Atomes tho it be hardly conceavable is not very uneasy evinceable." Using such examples as the light from colored glass, odoriferous effluvia, and the parts of mites as examples, Boyle shows that amazingly minute atoms are not really implausible despite their small size. The fact that red glass from a medieval church can still dye "atomes" of light with its redness despite its great age indicates the vast quantity—and hence small size—of the red atoms that it contains. Similarly, perfumed leather retains its scent for years, despite the fact that "each graine before it have quite forsaken the imbibing leather steemes away into Millions of fragrant Atomes."[28] As for the almost invisibly tiny mite, Boyle refers to microscopical observations that have revealed the various limbs on the animal. He remarks at the "multitude of Atomes" that must come together "to constitute the severall parts externall and internall necessary to make out this little Engine." He then asks us to consider "how unimaginably little must be the parts that make the haire upon the legs," again arguing for the unimaginably small size of atoms.[29] But none of these examples, importantly, require in themselves that we accept a particulate structure of matter. Why, after all, must the red light of the first example consist of atoms at all, not to mention the fragrant effluvia and insect hairs of the second and third? Rather, these examples assume that we have already adopted a corpuscular theory; they then try to explain away the commonsense objection that atoms cannot exist on account of their inconceivable minuteness. In short, this section must be understood as providing rhetorical support for the thesis already demonstrated in section 1 by Sennert's reduction to the pristine state.

In section 3, finally, Boyle focuses on the mysterious actions of the effluvia emitted by porous bodies. This section is by far the longest of the three parts and shows that even in the early 1650s, Boyle was intent

28. Boyle, *Atomicall Philosophy*, in Hunter and Davis, *Works*, vol. 13, p.228.
29. Boyle, *Atomicall Philosophy*, in Hunter and Davis, *Works*, vol. 13, p. 229.

on explaining the marvelous powers attributed to scholastic occult qualities in atomistic terms. He introduces this final section with the following words: "[T]hat there are Effluvia or steemes of Atomes issuing out of all bodyes seems not improbable nor that these Effluvia are extreamely subtle." Again, as in section 2, Boyle is not adducing experimental evidence here for the existence of atoms. Instead, he is assuming that atoms do exist and then using them to explain a host of remarkable phenomena. The fact that hunting dogs can detect the scent of a deer or partridge long after the animal has passed by does not of course force anyone to believe in the existence of atoms. After all, the effluvia emitted by the animal could just as well be a thin, continuous body like the pneuma of the ancient Stoics as a collection of discrete corpuscles. But if we suppose that these effluvia are really corpuscular in structure, then the particles out of which they are made must again be very tiny, since they pass continually and imperceptibly out of the pores of bodies.[30]

It is clear, then, that Boyle himself considered only the examples at the beginning of the *Atomicall Philosophy* to argue directly for the existence of atoms. And here only the chymical reductions to the pristine state provide experimental evidence for the claim "that there are Atomes," since only these examples argue for the semipermanence of the material corpuscles. It is precisely this semipermanence that makes them "atomic" as opposed to being just any bits of matter. The second and third sections, to reiterate, argue about the properties of corpuscles that are now assumed to be atomic on the basis of the first argument. Such properties include, above all, the facts that atoms are very small and that they steam out of other bodies.

If we now turn briefly to Boyle's *Sceptical Chymist*, printed in 1661 but composed in the final years of the 1650s, we shall see that Boyle himself was not long satisfied with the idea that the mineral acids dissolved materials into particles that were, strictly speaking, atomic. Nonetheless, the modifications that Boyle worked here on his earlier interpretation of Sennert's experiment can already be found in Sennert's own work. In other words, what we shall see in *The Sceptical Chymist* is largely a reinterpretation rather than a rejection of Sennert's chymical atomism. Let us proceed directly to the important part 1 of *The Sceptical Chymist*,

---

30. Here I disagree with Clericuzio, *Elements, Principles, and Corpuscles*, p. 119, who sees Boyle's comments on effluvia as a "proof" for the existence of atoms. This would presuppose an unwarranted degree of naiveté on Boyle's part.

where Boyle lays out his own propositions. These are the very principles of the mechanical philosophy. He begins with an assertion:

> *It seems not absurd to conceive that at the first Production of mixt Bodies, the Universal Matter whereof they among other Parts of the Universe consisted, was actually divided into little Particles of several sizes and shapes variously mov'd.*
> This (sayes *Carneades*) I suppose you will easily enough allow. For besides that which happens in the Generation, Corruption, Nutrition, and wasting of Bodies, that which we discover partly by our *Microscopes* of the extream little-nesse of even the scarce sensible parts of Concretes; and partly by the Chymical Resolutions of mixt Bodies, and by divers other Operations of Spagyrical Fires upon them, seems sufficiently to manifest their consisting of parts very minute and of differing Figures. And that there does also intervene a various local Mo-tion of such small Bodies, will scarcely be denied; whether we chuse to grant the Origine of Concretions asign'd by *Epicurus*, or that related by *Moses*. For the first, as you well know, supposes not only all mixt Bodies, but all others to be produc'd by the various and casual occursions of Atomes, moving themselves to and fro by an internal Principle in the Immense or rather infinite *Vacuum*. And as for the inspir'd Historian, He, informing us that the great and Wise Author of Things did not immediately create Plants, Beasts, Birds, &c., but produc'd them out of those portions of the pre-existent, though created, Mat-ter, that he calls Water and Earth, allows us to conceive, that the constituent Particles whereof these new Concretes were to consist, were variously moved in order to their being connected into the Bodies they were, by their various Coalitions and Textures, to compose.[31]

If we look at the text immediately following the italicized part of the proposition, the reader cannot help but be struck by the empirical weak-ness of Boyle's case. He mentions a number of examples from the world of the senses, such as the wasting or diminution of bodies and the way they appear under a microscope, to argue for "their consisting of parts very minute, and of differing Figures." But this is by no means evidence for a claim that such minute parts consist of a universal matter or that this ever had to be divided, even if Boyle is careful to couch his be-lief in terms of avoiding an absurdity rather than asserting a certainty.[32]

---

31. Boyle, *Sceptical Chymist*, in Hunter and Davis, *Works*, vol. 2, pp. 229-230.
32. Lest the reader accuse me of uncharitability here, it is important to be clear about what exactly Boyle wishes to assert. His argument, in short, is that God first created matter as a sort of lump, and only after this divided it into corpuscles while also imparting motion to it. This is more clearly spelled out in Boyle's contemporaneous *Usefulness of Natural Philosophy*, part 1, essay 4 (printed in 1663 but composed by about 1660, in Hunter and Davis, *Works*, vol. 3, p. 253), where Boyle is both criticizing the atheistic implications of Epicurean atomism and defending his own corpuscular theory: "Indeed, that the various coalitions of Atoms, or at least small Particles

Indeed, Boyle is clearly aware of the tenuous character of his avowal, for he immediately jumps to a sort of scholastic "proof by authority," invoking the word of Epicurus and Moses. The incessant movement of Epicurean atoms or the Mosaic composition of animals and plants from existing water and earth equally serve Boyle's purpose of showing the need for the transfer (and hence motion) of material corpuscles.

He then passes to his second proposition, which is the one that directly concerns us.

> Neither is it impossible that of these minute Particles divers of the smallest and neighbouring ones were here and there associated into minute Masses or Clusters, and did by their Coalitions constitute great store of such little primary Concretions or Masses, as were not easily dissipable into such Particles as compos'd them.[33]

Here we encounter something that was strikingly absent from Boyle's *Atomicall Philosophy*, namely, his new claim that the *minima* or *prima naturalia*, the smallest existing corpuscles, combine to form moleculelike aggregates of a semipermanent nature. Later in *The Sceptical Chymist* Boyle refers to these enduring aggregate particles as "*Prima Mista* or *Mista Primaria*."[34] In fact, this claim and its attendant terminology is already found clearly expressed in Sennert's *Hypomnemata physica*, which states that the smallest atoms are the corpuscles making up the four elements, fire, air, water, and earth. These combine to form what Sennert calls *prima mixta*, which are "minima of their own genus."[35] Like Boyle, Sennert treats the *prima mixta* as operationally atomic— they cannot be divided further by normal operations of man or nature.

---

of Matter, might have constituted the World, had not been perhaps a very absurd Opinion for a Philosopher, if he had, as Reason requires, suppos'd that the great Mass of lazy Matter was Created by God at the Beginning, and by Him put into a swift and various motion, whereby it was actually divided into small Parts of several Sizes and Figures, whose motion and crossings of each other were so guided by God, as to constitute, by their occursions and coalitions, the great inanimate parts of the Universe, and the seminal Principles of animated Concretions." For a probing study of Boyle's thoughts on seminal principles, see Anstey, "Boyle on Seminal Principles," pp. 597–630, especially p. 606.

33. Boyle, *Sceptical Chymist*, in Hunter and Davis, *Works*, vol. 2, p. 230.

34. Boyle, *Sceptical Chymist*, in Hunter and Davis, *Works*, vol. 2, p. 296.

35. Sennert, *Hypomnemata physica* (1636), pp. 107–108: "Sunt enim secundo alterius, praeter elementares, generis atomi, (quas si quis prima mista appellare velit, suo sensu utatur), in quae, ut similaria, alia corpora composita resolvuntur. Et omnino in mistione rerum naturalium, seu quae fit in non viventibus, corpora, e quibus mista constant, ita in exiguas partes confringuntur, & comminuuntur, ut nullum seorsim, & per se agnosci possit. In omnibus etiam fermentationibus

Hence Boyle is essentially reinterpreting a concept that he had elided or misunderstood at the time of writing his *Atomicall Philosophy*. In the earlier text he spoke only of "atoms" and did not consider the hierarchical nature of the compositional theory that Sennert had reworked from Geber and other chymical sources. By the time of the writing of *The Sceptical Chymist*, Boyle had come to realize that Sennert used the term "atom" in a loose, operational sense and that there were different genera or orders of such atoms, according to their stage of composition. His growing preference for other terms instead of "atom" probably reflects a dislike of the seeming imprecision in this usage (since the term *atomos* originally meant "indivisible" in an unqualified sense), as well as an attempt to avoid an explicit association with Epicurus.[36] At any rate, Boyle follows his second proposition with a long experimental justification for the existence of his semipermanent aggregate corpuscles.

> To what may be deduc'd, in favour of this Assertion from the Nature of the Thing it self, I will add something out of Experience, which though I have not known it used to such a purpose, seems to me more fairly to make out that there May be Elementary Bodies, then the more questionable Experiments of Peripateticks and Chymists prove that there Are such. I consider then that Gold will mix and be colliquated not only with Silver, Copper, Tin and Lead, but with Antimony, *Regulus Martis* and many other Minerals, with which it will compose Bodies very differing both from Gold, and the other Ingredients of the resulting Concretes. And the same Gold will also by common *Aqua Regis*, and (I speak it knowingly) by divers other *Menstruums* be reduc'd into a seeming Liquor, in so much that the Corpuscles of Gold will, with those of the *Menstruum*, pass through Cap-Paper . . . and many other wayes may Gold be disguis'd, and help to constitute Bodies of very differing Natures both from It and from one another, and nevertheless be afterward reduc'd to the self-same Numerical, Yellow, Fixt, Ponderous and Malleable gold it was before its commixture. Nor is it only the fixedest of Metals, but the most fugitive, that I may employ in favour of our Proposition: for Quicksilver will with divers Metals compose an *Amalgam*, with divers *Menstruums*, it seems to be turn'd into a Liquor, with *Aqua fortis* it will be brought into either a red or white Powder or precipitate, with Oyl of Vitriol into a pale Yellow one, with Sulphur it will compose a blood-red and volatile Cinaber, with some Saline Bodies it

---

& digestionibus ac coctionibus, quae vel a natura, vel ab arte fiunt, nihil aliud agitur, quam ut ad minima redigantur, & ea sibi arctissime uniantur."

36. For Boyle's ambivalent feelings with regard to Epicureanism, see Antonio Clericuzio, "A Redefinition of Boyle's Chemistry and Corpuscular Philosophy," *Annals of Science* 47 (1990): 561–589; see pp. 571–573.

will ascend in form of a Salt . . . And yet out of all these exotick Compounds, we may recover the very same running Mercury that was the main Ingredient of them, and was so disguis'd in them. Now the Reason (proceeds *Carneades*) that I have represented these things concerning Gold and Quicksilver, is, That it may not appear absurd to conceive, that such little primary Masses or Clusters, as our Proposition mentions, may remain undissipated, notwithstanding their entring into the composition of various Concretions, since the Corpuscles of Gold and Mercury, though they be not primary Concretions of the most minute Particles of matter, but confessedly mixt Bodies, are able to concurre plentifully to the composition of several very differing Bodies, without losing their own Nature or Texture, or having their cohaesion violated by the divorce of their associated parts or Ingredients.[37]

As we can see, this complicated-sounding series of reactions consists merely of reductions to the pristine state of gold and mercury. In the case of gold, Boyle says that one can alloy it with silver, copper, tin, lead, antimony, or other metals and then dissolve the gold out of the alloy, using the mixture of nitric and hydrochloric acid called aqua regia. As in the earlier experiment with silver, Boyle says that the solution will pass through cap paper, and yet the very same gold can be regained by reducing the solid acquired after its removal from the solution. A series of reactions with mercury then follows, the point being that no matter how much change is induced, the mercury can be regained intact. This passage too is probably inspired by a parallel *locus* in Sennert, but here I wish to pass to the importance of these reductions to the pristine state for Boyle's developed mechanical philosophy.[38]

If we compare the empirical evidence for the first and second propositions in *The Sceptical Chymist*, we find that the two enunciations wield a surprising disparity in their demonstrative force. In the first proposition, Boyle found it necessary to resort to the revealed wisdom of Moses as

37. Boyle, *Sceptical Chymist*, in Hunter and Davis, *Works*, vol. 2, pp. 230–231.
38. The Sennertian source is found in the 1636 *Hypomenemata physica* on pp. 114–115: "[I]f gold and silver be melted together, the atoms of gold and silver are so [thoroughly] mixed *per minima* that the former cannot be distinguished from the latter by the senses. Meanwhile both preserve their forms intact, which appears from this—that if aqua fortis be poured on that lump [of alloy], the silver dissolves and passes into the solution, but the gold remains in the form of a powder. The dissolved silver, if it be precipitated, [also] sinks down in the form of a powder. Each of these powders, if fused separately, returns to its original gold or silver. Thus if quicksilver be sublimed, precipitated, dissolve into a water, and assume other external mutations which come about in accordance with the diverse mixture of the atoms into which it is resolved with other [ingredients], nonetheless it always retains its own essential form, and is easily separated from the bodies with which it is mixed, and it is [thus] reduced into its original [form] of running mercury."

transmitted by the Bible. What other evidentiary basis did he have for claiming that an originally static and homogeneous mass of matter underwent division at the hand of God? In the second proposition, however, he was able to draw on the evidence of the reduction to the pristine state. As in the *Atomicall Philosophy*, Boyle's dissolution of gold was followed by a filtration that indicated the smallness of the metal's constituent particles, and the reacquisition of the intact metal revealed that its particles had remained undivided in the solution. In other words, this simple reversible reaction gave Boyle empirical verification of his claim that there are in nature semipermanent corpuscles, just as it had for Sennert. But for Boyle, this was only the beginning of the story, for the real project of his maturity lay not only in this claim but also in his oft-stated goal of reducing "particular qualities"—the phenomena of the sensible world— to mechanical causes. As we will show, this mechanical reductionism would rely heavily on the experimental strategies pioneered by Sennert, though now revised to reveal not only the particulate structure of matter but the ways in which that structure could be altered to engender qualitative change. Before we can proceed in this exposition, however, it will be necessary to consider Boyle's characterization of the mechanical philosophy itself.

THE MECHANICAL PHILOSOPHY AND THE ROLE OF EXPERIMENT

The "mechanical philosophy" has never been an unproblematic category. E. J. Dijksterhuis already pointed out a number of senses in which "mechanical" could be taken in his *Mechanization of the World Picture* of 1961, and Richard Westfall's *Construction of Modern Science* is built on the distinction between "mechanisms and mechanics."[39] One could readily devote a book to the different definitions that historians and philosophers have applied—implicitly or explicitly—to mechanism and the mechanical philosophy. My goal, however, is not to arrive at an essentialist picture of "mechanism" taken in a transhistorical sense. Instead, I hope to provide some modest clarification of the relationship that Robert Boyle conceived to exist between chymistry and the mechanical philosophy, the latter term being one that he surely did more than any other figure

39. E. J. Dijksterhuis, *The Mechanization of the World Picture* (Princeton: Princeton University Press, 1961), pp. 3–5. For further clarification, see the discussion of Dijksterhuis's views on mechanism in Floris Cohen, *The Scientific Revolution: A Historiographical Inquiry* (Chicago: University of Chicago Press, 1994), pp. 66–68; and Richard Westfall, *The Construction of Modern Science: Mechanisms and Mechanics* (Cambridge: Cambridge University Press, 1977) pp. 1–2 and passim.

to popularize. Only if we have a fairly distinct idea of what Boyle meant by the mechanical philosophy will it be possible to ascertain the full contours of his relationship to Sennert and earlier chymistry in general.

To this end, it is important to point to at least three common notions of "mechanism" that historians commonly import when considering Boyle and other premodern natural philosophers. The first is the idea that a given entity, or to use a terminology that is closer to that of the seventeenth century, a "body" can be explained in terms of its purely material parts, which in turn are characterized only by a short list of qualities capable of geometrical description, such as size, shape, position, arrangement, and perhaps motion (insofar as motion can be considered as a transposition from place to place).[40] This way of characterizing things belongs to the larger genus that Ernan McMullin dubbed "structural explanations."[41] Clearly it attempts to reduce the qualitative multiplicity of the phenomenal world to a tightly delimited set of structural factors. As we will see, this understanding of "mechanism," though seemingly innocuous, requires that the historian bent on explaining the idea carefully locate the parts of the structure that are intended to serve as explanatory tools. To give an arbitrary example, it would be one thing to explain the strength of a floor based on the spans of its joists and quite another to consider the molecules composing the joists, even though both explanations make an appeal to structure. As we will see, confusion regarding the scope of Boyle's structural explanations, particularly when it comes to his distinction between *prima naturalia* and *prima mista*, seriously mars some current historical reconstructions of his mechanism.

A second idea that is commonly brought to bear on seventeenth-century "mechanism" concerns the transmission of motion from body to body or from part to part. Thus Andrew Pyle, in his recent study of atomism, distinguishes explicitly between what he calls "reductionism" and "mechanism." By the former, he means the attempt to reduce the

---

40. I intentionally avoid the use of the terms "primary qualities" and "secondary qualities" here. As Peter Anstey points out in his recent study of Boyle's philosophy, these terms had a specific sense in scholastic natural philosophy that was quite different from Boyle's. In addition, Boyle himself employed the terms in ways that do not always correspond to the modern usage, which is descended from John Locke. See Anstey, *The Philosophy of Robert Boyle* (London: Routledge, 2000), pp. 21–30.

41. McMullin, "Structural Explanation," pp. 139–147. I say "larger genus" in part because Mc-Mullin does not limit "structure" to a body's *material* parts but defines it as "a set of constituent entities or processes and the relationships between them" (p. 139).

phenomena of the world to the structural factors that McMullin high-lighted. To Pyle, "mechanism" means something quite different, namely, a commitment to the idea that motion can only be transmitted by direct contact between bodies or parts of bodies as opposed to being induced by action at a distance. Pyle also introduces a third factor, however, claiming that the mechanical philosophy (a term that he applies to Descartes and other thinkers as well as Boyle) entails an adherence to nonteleological explanations.[42] In this Pyle is no doubt influenced by the longstanding tradition among historians of philosophy of contrasting teleology and mechanism, especially when discussing the ancient response to atom-ism.[43] Hence we have at least three distinct notions comprehended by historians under the term "mechanism"—structural reductionism, denial of action at a distance, and a rejection of final causes—each of them potentially vying for consideration as a sufficient definition but often combined.

Severe problems can arise when historians try to apply these concepts of "mechanism" to the mechanical philosophy, which was, after all, a seventeenth-century actor's category. The idea that the mechanical phi-losophy entailed a necessary opposition to teleology can be eliminated *ab initio*. Contemporary scholarship has revealed that Boyle, surely a prime exponent of the mechanical philosophy, was, if anything, a moderate supporter of teleological explanation.[44] In addition, as we pointed out above, Pyle explicitly tries to distinguish "the reductionist programme" from the mechanical philosophy. Peter Anstey, in his own recent treat-ment of the mechanical philosophy, criticizes Pyle for this and correctly points out that reductionism was an essential element of the mechan-ical philosophy even if there were reductionists who did not subscribe to the idea that all motion had to be transmitted by bodily contact.[45] More significant for our purposes is the fact that Boyle's mechanical philosophy was certainly a reductionist program and yet it was not

42. Pyle, *Atomism and Its Critics*, pp. 506–508.

43. For an excellent critique of the view that mechanism and teleology are mutually exclusive, see Heinrich von Staden, "Teleology and Mechanism: Aristotelian Biology and Early Hellenistic Medicine," in *Aristotelische Biologie*, ed. Wolfgang Kullmann and Sabine Föllinger (Stuttgart: Franz Steiner, 1996), pp. 183–208.

44. Timothy Shanahan, "Teleological Reasoning in Boyle's *Disquisition about Final Causes*," in Hunter, *Robert Boyle Reconsidered*, pp. 177–192; James G. Lennox, "Robert Boyle's Defense of Teleological Inference in Experimental Science," *Isis* 74 (1983): 38–53.

45. Anstey, *Philosophy of Robert Boyle*, pp. 12–13, n. 5. For Anstey's treatment of the mechanical philosophy more generally, see pp. 1–4.

wedded to hard claims about "the springs of local motion," to use Pyle's phrase.[46] Although Pyle's attempt at defining mechanism and the mechanical philosophy does damage to the historical record, other historians suffer from the fact that they are forced to undermine their own definitions. A case in point is Antonio Clericuzio, whose recent *Elements, Principles and Corpuscles* begins by defining his terms as follows—"[A]ny reference to the strict mechanical philosophy" will mean, in Pyle-like fashion, "a theory of matter according to which matter is inert and all interactions in nature are produced by the impact of particles." Yet near the end of his book, Clericuzio unwittingly illustrates the inadequacy of his own definition when he finds it necessary to equate a "strictly mechanical" explanation with one based solely on the shape, size, and motion of particles of inert matter: impact has silently dropped out of the picture.[47] More importantly, since Boyle's beliefs fit neither of these strict notions of the mechanical, Clericuzio is forced to exclude him from the fold of proper mechanical philosophers despite the fact that the British corpuscularian spent the better part of his life trying to justify a set of scientific beliefs that he himself dubbed "the mechanical philosophy." If anyone should be a mechanical philosopher, it is Boyle.[48]

Another difficulty in the approach of Pyle and Clericuzio, as well as many older authors, is their implicit reliance on Cartesianism in framing their definitions of the mechanical philosophy. Although Boyle was obviously influenced by Descartes, he was, at the same time, a self-styled follower of Bacon. As Rose-Mary Sargent has rightly argued, Boyle's scientific commitments lay more with the empirical verification of his corpuscular theory than they did with aprioristic Cartesian claims about the ultimate nature of matter and motion. Unlike Descartes, Boyle was both temperamentally and philosophically opposed to making

46. See Anstey, *Philosophy of Robert Boyle*, pp. 153–154. Anstey argues, I believe correctly, that "there is no prospect of unearthing a coherent physics in Boyle's works... Boyle was first and foremost a philosopher of the qualities, not motion, space or time."

47. Antonio Clericuzio, *Elements, Principles and Corpuscles: A Study of Atomism and Chemistry in the Seventeenth Century* (Dordrecht: Kluwer, 2000), pp. 7, 106.

48. As Anstey points out, Boyle did not himself create the term "mechanical philosophy." That honor may go to Henry More, who in responding to Descartes speaks of a "mechanick philosophy" (see Anstey, *Philosophy of Robert Boyle*, p. 12; and More, *The Immortality of the Soul* [London: William Morden, 1659], Preface, [b6^r], [b8^r], and passim. Nonetheless, it is Boyle, not More, who popularized the expression, so it seems rather anachronistic to deny to Boyle the status of a mechanical philosopher. Instead of making this move, historians such as Clericuzio should reexamine their own understanding of the meaning that Boyle attributed to the term.

categorical statements on the basis of first principles.[49] It is a peculiar irony, then, that Cartesian views about the strict reduction of phenomena to the primary qualities of inert matter or about the transmission of motion between bodies should be treated as defining elements of "the mechanical philosophy" when it was Boyle—rather than Descartes— who made use of that expression.

Partly as a result of viewing the mechanical philosophy in Cartesian terms as a system derived from first principles, writers such as Clericuzio and more recently Alan Chalmers have stressed the mutually divergent character of Boyle's chymistry and his mechanical philosophy. In a widely cited article published in 1990, Clericuzio argues that Boyle's chymistry occupied a domain quite distinct from his mechanical philosophy and was not subject to its strictures.[50] In this chapter, I will argue to the contrary that Boyle's chymistry frequently appeared in his works precisely in order to demonstrate the mechanical operations of corpuscles, even in the instances that Clericuzio considers to be the most compelling for his case. Clericuzio's approach is built on Boyle's distinction between the *prima naturalia* or *minima naturalia*—the first or smallest particles that cannot be divided further, and the aggregate corpuscles, which begin with the *prima mista* (i.e., *prima mixta*—first compounded bodies) and continue upwards to form more and more complex particles ("decompounded" bodies).[51] The corpuscles of lowest stage (*prima*

49. Maries Boas Hall already pointed out Boyle's emphasis on Bacon over Descartes in her influential "Establishment of the Mechanical Philosophy," pp. 460–464. See also Sargent, *Diffident Naturalist*, pp. 35–41.

50. Clericuzio, "Redefinition," p. 563. Clericuzio's *Elements, Principles, and Corpuscles*, pp. 103–148, presents the same argument as the 1990 article, but in somewhat attenuated form. See in particular p. 145, where Clericuzio reaffirms his "non-mechanical" explanation of Boyle's indicator test involving mercuric chloride and sulfuric acid. Since the article is more detailed than the book, I will rely on the former for my critique of Clericuzio's position.

51. Boyle, *The Sceptical Chymist*, in Hunter and Davis, *Works*, vol. 2, pp. 296–297, 347; *Forms and Qualities*, in Hunter and Davis, *Works*, vol. 5, p. 326; "The History of Particular Qualities," in Hunter and Davis, *Works*, vol. 6, 270–271, 274–275; "Experiments, and Notes, about the Mechanical Origine and Production of Volatility," in Hunter and Davis, *Works*, vol. 8, p. 425; *Experiments and Notes about the Producibleness of Chymicall Principles*, in Hunter and Davis, *Works*, vol. 9, p. 114. Throughout the present book I will use the term "aggregate corpuscle" to refer to Boyle's semipermanent clusters of *prima naturalia*, without regard to their level of composition. The significant thing for our purposes is the distinction between the *prima naturalia* and their aggregates, not the distinctions among the different levels of aggregation. Note that Boyle's term "decompounded" does not mean "uncompounded," but rather "twice compounded." "Decompounded" was before Boyle a grammatical term for words built from more than two other words, as in the Latin *superexaltavit* (see *The Compact Edition of the Oxford English Dictionary* (Oxford: Oxford University Press, 1971), s.v. "decompound"). But the term "decompounded" had

naturalia) are endowed with only the essential characteristics of matter, which can be comprehended under the three generic categories of size, shape, and motion (or its absence). Boyle called these characteristics "mechanical affections," among which he also frequently included "texture" or "contexture," which referred to the structural attributes of a group of *prima naturalia* or of higher-order corpuscles.[52] At other times, Boyle used the term "catholic affections" for size, shape, and motion, since these qualities were shared by all matter; he did not usually consider "texture" to be a catholic affection, since it was a collective property rather than a characteristic of a single ultimate particle.[53] In the case of the aggregate corpuscles, as opposed to the *prima naturalia*, Clericuzio maintains that Boyle did not attempt to derive their perceptible properties and powers from the "mechanical affections." In Clericuzio's analysis Boyle was content to treat the *prima mista* and more complex aggregates as self-subsistent "chemical corpuscles" endowed with a host of nonmechanical properties providing them with their specific identities and their ability to react among themselves. Hence Clericuzio says that Boyle "regarded chemistry as a discipline independent from mechanics" and therefore "his chemistry can be described as corpuscular, not mechanical."[54]

---

already been absorbed into discussions of matter theory by the time of Heinrich Cornelius Agrippa's *De occulta philosophia*, if not earlier. See Agrippa, *De occulta philosophia libri tres*, ed. V. Perrone Compagni (Leiden: Brill, 1992), p. 91.

52. As in Boyle, "About the Excellency and Grounds of the Mechanical Hypothesis," in Hunter and Davis, *Works*, vol. 8, p.111, where he speaks of "the union of insensible particles in a convenient Size, Shape, Motion or Rest, and Contexture; all which are but Mechanical Affections of convening Corpuscles." See also "About the Excellency and Grounds of the Mechanical Hypothesis," p. 105, as well as "Of the Imperfection of the Chymist's Doctrine of Qualities," in Hunter and Davis, *Works*, vol. 8, pp. 401–402; "Experiments and Notes about the Mechanical Origine and Production of Volatility," in Hunter and Davis, *Works*, vol. 8, p. 431; *Colours*, in Hunter and Davis, *Works*, vol. 4, p. 99; *Sceptical Chymist*, in Hunter and Davis, *Works*, vol. 2, p. 356 (where Boyle explicitly equates "structure" and "texture"), and *Forms and Qualities*, in Hunter and Davis, *Works*, vol.5, pp. 306, 310, and passim.

53. There are exceptions, of course, as in a passage from *Forms and Qualities* where Boyle includes texture among the "more Catholick Affections of Bodies," but here he is speaking comparatively, and relating texture to colors, tastes, and other secondary qualities. See *Forms and Qualities*, Hunter and Davis, *Works*, vol. 5, p. 319.

54. Clericuzio, "Redefinition," p. 563: "In addition, at a closer investigation of Boyle's chemical published works and manuscripts, it becomes apparent that he was far from subordinating chemistry to mechanical philosophy, since he did not explain chemical phenomena by immediate and direct recourse to the mechanical affections of particles. As a matter of fact, he regarded chemistry as a discipline independent from mechanics. He explained chemical phenomena

Clericuzio's approach has been adopted and expanded more recently by Alan Chalmers, who identifies two divergent programs in Boyle's work, one of which he calls "science" (mostly chemistry and pneumatics) and the other "the mechanical philosophy." According to Chalmers, Boyle's mechanical philosophy consisted of the attempt to reduce all natural phenomena to the mechanical or catholic affections of the *prima naturalia* and the spatial relationships among them.[55] Since according to Chalmers this project was riddled with inconsistencies and gratuitous assumptions, Boyle's real successes in the realm of chemistry and pneumatics occurred "in spite of, rather than because of" his allegiance to the mechanical philosophy.[56]

The stark distinction that Clericuzio and Chalmers draw between Boyle's corpuscular chemistry and his mechanical philosophy cannot hold, however.[57] One will certainly agree that Boyle was deeply interested in chymistry for its own sake as also for its Baconian (and Helmontian) usefulness—he was, after all, an aspirant to such chymical *Arcana Maiora* as the philosophers' stone and the alkahest.[58] It is also true that Boyle does not generally try to impose a strict reductionism on chymical phenomena by leading them back to the catholic affections of

---

in terms of corpuscles endowed with chemical, rather than mechanical properties. Accordingly, his chemistry can be described as corpuscular, not mechanical."

55. Alan Chalmers, "The Lack of Excellence of Boyle's Mechanical Philosophy," *Studies in History and Philosophy of Science* 24 (1993): 541–563; see pp. 543–544: "Boyle's mechanical or corpuscular hypothesis is spelt out in most detail in 'The Origin of Forms and Qualities According to the Corpuscular Philosophy' although the general features of it are described by Boyle in many of his works. A key feature of it is its reductionist character. All the phenomena of the material world are to be reduced to the action of matter and motion . . . In the following I use the term 'corpuscle' to refer to minima [i.e., Boyle's *prima naturalia*] only. Each corpuscle possesses a determinate shape, size, and degree of motion or rest . . . For Boyle the secondary qualities are to be reduced to, that is, explained in terms of, the primary ones. More specifically, all the phenomena of the material world are to be explained in terms of the shapes, sizes, and motions of corpuscles together with the spatial arrangements of those corpuscles amongst themselves."

56. Alan Chalmers, "The Lack of Excellence of Boyle's Mechanical Philosophy," *Studies in History and Philosophy of Science* 24 (1993): 541–563; see p. 541.

57. Chalmers's 1993 paper has in fact proven to be controversial. Recently he has engaged in a three-part exchange with Andrew Pyle and Peter Anstey where he attempts to clarify his views but does not back down. See Chalmers, "Experimental versus Mechanical Philosophy in the Work of Robert Boyle: A Reply to Anstey and Pyle," *Studies in History and Philosophy of Science* 33 (2002): 191–197. See also Anstey, "Robert Boyle and the Heuristic Value of Mechanism," pp. 161–174; and Pyle, "Boyle on Science and the Mechanical Philosophy: A Reply to Chalmers," pp. 175–190.

58. Principe, *Aspiring Adept*, passim.

the *prima naturalia*, but it does not follow from this that his chymistry is not mechanical. Indeed, the vast majority of Boyle's chymical examples in *Forms and Qualities*, *Colours*, and other works on natural philosophy are intended to demonstrate that changes in the texture (a mechanical affection) of chymical corpuscles are the cause of the qualitative changes traditionally attributed to Aristotlelian forms or to the chymical principles.[59] The fact that Boyle does not attempt to reduce all phenomenal change to the level of the *prima naturalia* or initial particles does not mean that his chymical explanations are not mechanical, since the aggregate corpuscles are also endowed with mechanical affections having explanatory force.

Clericuzio considers explanations in terms of the "the recess of some particles and the access of some others" as being nonmechanical if the particles involved had properties other than the catholic affections of the *prima naturalia*.[60] But Boyle's explanations employing the association (access), dissociation (recess), and rearrangement of corpuscles that are endowed with enduring essential properties are closely related to Sennert's reductionist employment of *synkrisis*, *diakrisis*, and immutation, which we have already discussed. Why should we exclude structural explanations of this sort from the realm of the mechanical philosophy? In fact, it is clear that Boyle himself did not do so. The key piece of Clericuzio's evidence is, chymically speaking, a reduction to the pristine state, although not Sennert's classic dissolution and reduction of metals. The experiment is related by Boyle in *Colours*, where white mercury sublimate (mercuric chloride) is dissolved in water to form a colorless solution. When Boyle adds salt of tartar, an orange color appears, which can in turn be made to disappear by dropping in some oil of vitriol (sulfuric acid). Subsequently, Boyle offers what he calls the "Chymical reason" of the process. He asserts that the color results from the association of mercurial and saline particles, and when these are caused to dissociate from one another, the color again disappears. The cycle of clear solution—orange solution—clear solution, viewed in terms of *synkrisis* followed by *diakrisis*, thus fulfils the requirement of a Sennertian *reductio in pristinum statum*. But Boyle then distinguishes this "Chymical reason" from a "truly Philosophical or Mechanical" explication, saying that the latter, if attainable, would tell us not only why a color change occurred,

59. Oddly, Clericuzio seems to view texture as a nonmechanical property, despite its clear appeal to structure. See Clericuzio, *Elements, Principles, and Corpuscles*, p. 148.
60. Clericuzio, "Redefinition," pp. 578–579.

but "why the Particles of the *Mercury*, of the Tartar, and of the Acid Salts convening together, should make rather an Orange Colour than a Red, or a Blew, or a Green."[61] Because Boyle labels his own interpretation of the experiment a "Chymical reason" rather than a "truly Philosophical or Mechanical" one, Clericuzio concludes that Boyle universally viewed explanations in terms of the access and recess of chymical corpuscles as being nonmechanical. Unfortunately, Clericuzio has overlooked a passage that severely undermines his interpretation, for in it Boyle explicitly says that the production and disappearance of the orange color reveal its mechanical origin. Boyle introduces the experiment as follows:

> The Experiment I am now to mention to you, *Pyrophilus*, is that which both you, and all the other *Virtuosi* that have seen it, have been pleas'd to think very strange; and indeed of all the Experiments of Colours, I have yet met with, it seems to be the fittest to recommend the Doctrine propos'd in this Treatise, and to shew that we need not suppose, that all Colours must necessarily be Inherent Qualities, flowing from the Substantial Forms of the Bodies they are said to belong to, since by a bare Mechanical change of Texture in the Minute parts of Bodies, two Colours may in a moment be Generated quite *De novo*, and utterly Destroy'd. For there is this difference betwixt the following Experiment, and most of the others deliver'd in these Papers, that in this, the Colour that a Body already had, is not chang'd into another, but betwixt two Bodies, each of them apart devoid of Colour, there is in a moment generated a very deep Colour, and which if it were let alone, would be permanent; and yet by a very small Parcel of a third Body, that has no Colour of its own, (lest some may pretend I know not what Antipathy betwixt Colours) this otherwise permanent Colour will be in another trice so quite Destroy'd, that there will remain no foot-steps either of it or of any other Colour in the whole Mixture.[62]

Clearly Boyle views the experiment as demonstrating a basic fact of the mechanical philosophy, namely, that changes in color can be induced and removed "by a bare Mechanical change of Texture in the Minute parts of Bodies." More revealing than this fact alone is Boyle's use of the superlative "fittest." The experiment by which an orange liquid is produced from two that are colorless and then removed is not merely convincing; it is "the fittest" known to him for its destruction of the scholastic theory that colors depend on substantial forms. At the same time, however, the experiment is the "fittest to recommend the Doctrine propos'd in this Treatise." To what doctrine is Boyle alluding? Surely the

61. Boyle, *Colours*, in Hunter and Davis, *Works*, vol. 4, p. 152.
62. Boyle, *Colours*, in Hunter and Davis, *Works*, vol. 4, p. 150.

doctrine under discussion must be the mechanical philosophy. After all, as Boyle tells us earlier in *Colours*, his goal for the book is to explain the generation of colours in bodies "by Intelligible and Mechanical principles."[63] In particular, Boyle wants to demonstrate by experiment that the modification of light to produce different colors "depends upon the continuing or alter'd Texture of the Object." Expounding on this further, Boyle says that it is not only the difference of shape in the constituent corpuscles that causes colors to vary but also their varying sizes, close or loose order, mixing of different particle types, size and shape of the fissures between particles, and also the "Beams of Light," which although "subtil Bodies," have their own roughness.[64] It is obviously a thoroughly mechanical explanation of color that forms "the Doctrine propos'd in this Treatise." Hence it is no surprise when Boyle asserts that the experiment for producing an orange solution from two clear ones depends on a mechanical change of texture in the "Minute parts" of the bodies involved.

So what, then, does Boyle mean when he says that the access and recess of chymical corpuscles provides a "Chymical reason" rather than "a "truly Philosophical or Mechanical" one? First, since the processes involve mercury (in the form of the mercury sublimate) and "salts" (an "alkaline salt" in case of the oil of tartar and an "acid salt" in that of oil of vitriol) Boyle may have feared that he could be accused of employing the chymical principles of the Paracelsians as explanatory tools. But there is more to Boyle's reticence than this. Clearly Boyle's dissatisfaction with his "chymical" explanation lies mainly in its failure to explain the generation of one color rather than another. But in reality, we will find such an explanation of individual colors nowhere in Boyle's book. Boyle's theory of colors relied on the assumption that whiteness was a phenomenon of reflection, brought about when the reflecting particles were so small that the images borne by each were too small to be made out.[65] The production of colors, on the other hand, he believed to result from a modification of light involving change of texture, but he had no mechanical explanation for the generation of one primary color rather than another. It is not only in the realm of chymical corpuscles, but in that of the mechanical philosophy as a whole that Boyle failed to arrive

63. Boyle, *Colours*, in Hunter and Davis, *Works*, vol. 4, p. 61.
64. Boyle, *Colours*, in Hunter and Davis, *Works*, vol. 4, pp. 35–37.
65. Boyle, "History of Particular Qualities," in Hunter and Davis, *Works*, vol. 6, pp. 280–281; "Reflections upon the Hypothesis of Alcali and Acidum," in Hunter and Davis, *Works*, vol. 8, p. 411; and Boyle, *Colours*, in Hunter and Davis, *Works*, vol. 4, p.37.

at specific mechanisms for the production of individual colors. Indeed, he admitted as much himself, for in part 1 of *Colours* he stated that his immediate goal was only to provide "an *Apparatus* to a sound and comprehensive [*sic!*] Hypothesis," rather than to determine "the making out of the Generation of Particular Colours."[66] When Boyle complains in his mercurial experiment that he has not yet arrived at a "truly philosophical or mechanical" explanation of the orange formed from mercury sublimate, then, he is merely acknowledging the preliminary character of his research and his well-known reluctance to commit himself to precise mechanical models.[67] It does not follow, of course, that he thinks his explanation in terms of the association and dissociation of aggregate corpuscles not to be "mechanical" in an unqualified sense. How could it not be mechanical, given that it involves a change of texture, one of the "mechanical affections"? Rather it is not "truly" mechanical in the sense that it fails to provide the full explanatory precision to which Boyle aspires (and admittedly does not attain). It is "Chymical," on the other hand, to the extent that it appeals to the chymical properties of the corpuscles in order to explain the source of their mutual association and dissociation (not because it employs association and dissociation as explanatory tools). As if to confirm these points, Boyle reiterates once more a few lines after mentioning his "Chymical reason" that the experiment indeed reveals that color depends "upon the Texture resulting from the Convention of the several sorts of Corpuscles."[68]

Since we cannot restrict Boyle's mechanical philosophy to an attempt to explain natural phenomena purely in terms of *prima naturalia* having only the catholic affections, then how should we view it? Although one sense of the term "mechanical" in the seventeenth century was still that of "practical" or "artisanal," as in the Latin term *ars mechanica*, it is clear that Boyle's use of the term in the phrase "mechanical philosophy" was meant primarily to relate to mechanisms or machines.[69] Rather than merely taking it for granted that "mechanical qualities" are synonymous with the qualities of the initial particles, then, let us make an initial attempt to determine what it was about machines that led Boyle to apply

---

66. Boyle, *Colours*, in Hunter and Davis, *Works*, vol. 4, p. 58.
67. Sargent, *Diffident Naturalist*, pp. 105–106, 125.
68. Boyle, *Colours*, Hunter and Davis, *Works*, vol. 4, p. 152.
69. For the medieval "mechanical arts," see Peter Sternagel, *Die Artes Mechanicae im Mittelalter: Begriffs- und Bedeutungsgeschichte bis zum Ende des 13. Jahrhunderts*, Münchener Historische Studien, Abteilung Mittelalterliche Geschichte, Herausgegeben von Johannes Spürl, Band II (Kallmünz über Regensburg: Michael Lassleben, 1966).

their characteristics to his material theory. In this effort, a well-known passage from *Forms and Qualities* will prove useful.

> That then, which I chiefly aime at, is to make it Probable to you by Experiments (which I Think hath not yet beene done:) that allmost all sorts of Qualities, most of which have been by the Schooles either left Unexplicated, or Generally referr'd, to I know not what Incomprehensible Substantiall Formes, *may* be produced Mechanically, I mean by such Corporeall Agents, as do not appear; either to Work otherwise, then by vertue of the Motion, Size, Figure, and Contrivance of their own Parts, (which Attributes I call the Mechanicall Affections of Matter, because to Them men willingly Referre the various Operations of Mechanical Engines:) or to Produce the new Qualities exhibited by those Bodies their Action changes, by any other way, then by changing the *Texture*, or *Motion*, or some other *Mechanical Affection* of the Body wrought upon.[70]

Here, as in other passages, Boyle explicitly states that the term "mechanical" refers to the motion, size, figure, and contrivance—or mutual ordering, which he usually calls "texture"—of the parts that make up a body. Although there were many other features that Boyle found appealing about machines, it is clear that their ease of explanation in terms of the size, shape, and interaction of their parts was of paramount importance to his mechanical philosophy. Unlike substantial forms, whose precise nature even the scholastics viewed as escaping the senses and hence remaining forever "unknowable," machines and their parts had properties that were capable of easy visualization and comprehension. These properties or "affections" are "mechanical" because in the traditional discipline of mechanics, descending from the Hellenistic engineers, they are all that one needs to explain the operations of machines: the mechanical advantage to be supplied by a lever, for example, depends on its size and the position of the fulcrum rather than on whether it is made of bronze or wood. In the case of gears, on the other hand, the diameter of the wheels and their axes and the relation of the gears to one another will determine the transmission of power. Although the presence or absence of friction and the resistance of the device to breakage would obviously be conditioned by its material, it is the size, shape, and interrelation of its parts that determine its structural capability as a machine.[71] The same may be said of other simple machines—the function of

70. Boyle, *Forms and Qualities*, in Hunter and Davis, *Works*, vol. 5, p. 302.
71. A good example of this geometrical approach to machine design may be found in Hero's discussion of gears, for which see Héron d'Alexandrie, *Les mécaniques ou l'élévateur des corps lourds*, ed. and trans. by B. Carra de Vaux (Paris: Belles Lettres, 1988), pp. 217–226.

the pulley, inclined plane, screw, or windlass is in principle independent of its material composition. It is therefore unnecessary to invoke the traditional Aristotelian "primary qualities" of hot, cold, wet, and dry to explain their operation, nor must one resort to the presence or absence of the Paracelsian *tria prima*, mercury, sulfur, and salt. This advantage of mechanistic explanations is spelled out even more clearly in a passage from *The Sceptical Chymist*.

> And however Chymists boldly deduce such and such properties from this or that proportion of their component Principles; yet in Concretes that abound with this or that Ingredient, 'tis not alwayes so much by vertue of its presence, nor its plenty, that the Concrete is qualify'd to perform such and such Effects; as upon the account of the particular texture of that and the other Ingredients, associated after a determinate Manner into one Concrete (though possibly such a proportion of that ingredient may be more convenient than an other for the constituting of such a body.) Thus in a clock the hand is mov'd upon the dyal, the bell is struck, and the other actions belonging to the engine are perform'd, not because the Wheeles are of brass or iron, or part of one metal and part of another, or because the weights are of Lead, but by Vertue of the size, shape, bigness, and co-aptation of the several parts; which would performe the same things though the wheels were of Silver, or Lead, or Wood, and the Weights of Stone or Clay; provided the Fabrick or Contrivance of the engine were the same: though it be not to be deny'd, that Brasse and Steel are more convenient materials to make clock-wheels of than Lead, or Wood.[72]

In this passage, Boyle compares the "texture" of the ingredients in a chymical compound to the "coaptation" of the parts in a clock. The latter term means what we would call "fitting together" or perhaps even "structure." The working of a clock depends on the way that its parts fit together to produce an interactive structure, a mechanism. The parts must have certain characteristics, such as rigidity, but various different materials quite diverse in their other properties can supply that quality. The mechanical analysis of the clock takes such material properties for granted—they do not figure as components of the explanation. Hence the materials out of which the parts of the clock are made are irrelevant to a mechanical analysis of it, and in a certain way one could add that the same is true of the mechanical analysis of the physical world at large. For such a mechanical explanation to occur, there is no need to invoke a substrate having only the catholic affections, since the explanation itself renders material difference irrelevant. But problems begin to arise when

72. Boyle, *Sceptical Chymist*, in Hunter and Davis, *Works*, vol. 2, pp. 341–342.

the observer cannot see the parts that are supposed to be interacting in a mechanical way, as when they are invisibly small. What assurance do we have in such a case that the qualities perceived by us can be reduced to mechanical properties such as texture instead of flowing from a substantial form or emanating from one of the three Paracelsian principles? As we will see, Sennert's reduction to the pristine state reappears in Boyle's work as a response to this question.

But first let us return to the fallacious idea that in order for an explanation to count as "mechanical" in Boyle's system it had to employ the catholic affections of the smallest atoms—the *prima naturalia*. This does not follow from the passages just cited, nor does it emerge naturally from other Boylean treatments of the topic. All macrolevel bodies (not to mention machines proper) must obviously be endowed with mechanical properties. Similarly, the microlevel aggregate corpuscles must also have size and shape and be capable of being set into motion.[73] And if we look at the vast majority of Boyle's experimental demonstrations of the mechanical origins of qualities, it is precisely such aggregate corpuscles that usually come into discussion. The explosive quality of gunpowder, for example, is due, in Boyle's words, to "the Mechanical Characterization or *Stamp* of Matter" in the form of the "bare comminution and blending" of the particles of sulfur, saltpeter, and charcoal in it.[74] These are aggregate corpuscles distinguished by their chymical properties, not initial atoms having primary qualities alone. Nonetheless, it is the "mechanical texture" formed by the association of these finely ground particles that produces the explosiveness of gunpowder.[75] A multitude of similar explanations occur in Boyle's treatment of colors. If one scrapes off a small bit of black horn, for example, Boyle says it will appear white. Boyle explains that "this so great and sudden change is effected by a slight Mechanical Transposition of parts . . . the Effect proceeding only from a Local Motion of the parts." Although the bits of horn retain their chymical properties or "hornness" and are not reduced to atomic size, their transposition by local motion is viewed as a "mechanical" explanation of their change in color.[76]

A related experimental *modus operandi* that Boyle often used consisted of altering, producing, or destroying a sensible quality in a substance

73. This point has also been made by Anstey, "Robert Boyle and the Heuristic Value of Mechanism," p. 166.

74. Boyle, *Forms and Qualities*, in Hunter and Davis, *Works*, vol. 5, p. 369.

75. Boyle, *Forms and Qualities*, in Hunter and Davis, *Works*, vol. 5, p. 460.

76. Boyle, *Colours*, in Hunter and Davis, vol. 4, pp. 96–97.

by mechanical means. Once he had effected a sensible change mechanically, he could then argue (on the basis of parsimony and burden of proof) that mechanical modes of operation must be at the basis of that quality. The inducing of sensible change in the qualities of a substance by mechanical means, then, allowed Boyle to argue that such sensible qualities were mechanical in origin, regardless of the stage of composition of the particles concerned. The result of Boyle's characteristic type of experimentation was that the mechanical philosophy, despite its programmatic appeal to the catholic affections of the *prima naturalia*, was only partially reductionistic. Instead of appealing to the size, shape, and motion of the uncompounded and initial particles, Boyle frequently based his mechanical explanations on the access, recess, or transposition of unchanging aggregate corpuscles with chymical properties.[77] It does not follow from the fact that these explanations are not based on the catholic affections of the *prima naturalia* that they are not mechanical. The radical disjunction claimed by some scholars to exist between Boyle's mechanical philosophy and chymistry is in fact illusory. We are now in a position, then, to see how the corpuscular demonstrations pioneered by Daniel Sennert and adopted early on by the young Boyle as proofs for the existence of enduring corpuscles served the British virtuoso in his maturity as a means of support for the mechanical philosophy as a whole. The manifest traces of the German scholastic's matter theory on the mechanism of the fully developed Boyle can only come as a surprise to those who have been nurtured on the image of the British "naturalist" as the embodiment of a "new science" hewn from the rejection of Aristotelian natural philosophy.

77. See Boyle, *Forms and Qualities*, vol. 5, p. 399,where he describes "what great Changes may be made, even in Bodies scarce corruptible, by one or more of those three Catholick wayes of Natures working according to the Corpuscular Principles, namely, the Access, the Recess, and the Transposition of the minute Particles of Matter." On page 328, Boyle equates the combination and mutual separation of aggregate corpuscles with the *synkrisis* and *diakrisis* of the ancient atomists.

7

Boyle's Use of Chymical Corpuscles and the Reduction to the Pristine State to Demonstrate the Mechanical Origin of Qualities

Boyle's mechanical philosophy and his chymistry were closely integrated threads in his overall attempt to carry out the research program of Francis Bacon. Bacon himself was significantly influenced by alchemy in a number of ways, as various scholars have revealed.[1] But in addition to his borrowings from Bacon and others, Boyle's project, as we have now seen, also owed a major debt to Daniel Sennert and the tradition of corpuscular chymistry stretching back to Geber and other medieval alchemists. Boyle's work is full of experimental examples where he describes aggregate corpuscles undergoing mechanical change as a result of what Sennert would have called *synkrisis*, *diakrisis*, and immutation.

---

1. Bacon's debts to alchemy are manifold. Graham Rees has argued that Bacon's matter theory owed a significant debt to Paracelsus. See Rees, "Francis Bacon's Semi-Paracelsian Cosmology," *Ambix* 22 (1975): 81–101. Paolo Rossi thought that Bacon imbibed some of his technological optimism from alchemy and natural magic. See Rossi, *Francesco Bacone: Dalla Magia alla Scienza* (Bari: Laterza, 1957), pp. 54–62. I have made the argument that Bacon's emphasis on the ability of human art to replicate the products of nature rather than producing mere second-rate simulacra owes a heavy debt to the tradition of scholastic alchemy stretching back to the Middle Ages, as does his related concept of "Maker's Knowledge." See Newman, *Promethean Ambitions* (Chicago: University of Chicago Press, 2004), chapter 5.

In effect, Boyle treats his aggregate corpuscles as "minima of their own genus," like the particles that Sennert demonstrated to exist by means of his reductions to the pristine state. In some instances, Boyle's mechanical explanations could even seem to have been inspired directly by Sennert, as when the British "naturalist" explains the explosive power of gunpowder as resulting from an alteration in the texture of its ingredients.[2] Sennert had accounted for the same phenomenon by saying that it arose merely from the fine grinding and close juxtaposition of the particles of niter, sulfur, and charcoal in the gunpowder. In the following, we will see that Boyle relied on Sennert's reductions to the pristine state not merely to show that semipermanent corpuscles existed in nature but as part of a comprehensive strategy to reveal that most qualitative change was at basis mechanical.

Before proceeding down this path, however, it will be useful briefly to show how Boyle and Sennert differed from another important corpuscular theorist, Pierre Gassendi, in their emphasis on that chymical proof par excellence, the *reductio in pristinum statum*. The reader will recall that Boyle already invoked Gassendi's name in his juvenile essay *Of the Atomicall Philosophy* but without making much if any real use of Gassendi's ideas. Yet in his more mature works of the 1660s and later, Gassendi came to be a signal source of Boyle's microstructural explanations. As we will now see, however, in comparison to the British "naturalist" and his German predecessor, Gassendi made less use of this particular demonstration and failed, on the whole, to see its significance. This will serve to underscore the particularity of the chymical tradition shared by Boyle and Sennert.

It is well known that Gassendi had a strong appreciation of the analytical capabilities of chymistry, despite his engagement in a bruising attack on the English Rosicrucian enthusiast Robert Fludd in 1629.[3] At times Gassendi even endorsed the chymical doctrine that the principles salt and sulfur were responsible for taste and smell. Since Gassendi believed (like Sennert and Boyle) that the primordial atoms combined with one another

---

2. It must be noted, however, that Francis Bacon also had a similar explanation for the deflagrating action of gunpowder. Like Sennert, Bacon attributed the rapid expansion of gases to the action of niter trying to escape the fiery sulfur. Unlike Sennert, however, Bacon here makes no special point of the change in texture of the ingredients induced by their fine grinding. See Bacon, *Novum organum*, in James Spedding et al., *The Works of Francis Bacon* (London: Longman et al., 1860), vol. 4, p. 188.

3. Olivier René Bloch, *La philosophie de Gassendi: Nominalisme, matérialisme et métaphysique* (La Haye: Martinus Nijhoff, 1971), pp. 233–278.

to form compound corpuscles, which he called *moleculae* ("molecules"), it was possible for him to see the chymical principles as being higher-order corpuscles. Hence they were not incompatible with his atomism.[4] Gassendi also eagerly employed chymistry to show that new qualities could be induced where they had not existed before, and like Boyle, he often interpreted such changes in terms of textural alteration. An important example of this appears in Gassendi's early *Animadversiones in decimum librum Diogenis Laertii* (1649), a text that Boyle knew well.[5]

The experiment in question, which may have served as a partial inspiration to Boyle's own work on the color changes wrought by salt of tartar and oil of vitriol on a mercuric chloride solution (discussed in the previous chapter), comes within striking distance of being a reduction to the pristine state, and yet it remains something else. In brief, Gassendi places a handful of senna leaves in warm water, then adds oil of tartar (dissolved potassium carbonate, usually melted in the open air by hygroscopic action), and observes that the solution becomes red. He then asks himself rhetorically where the redness has come from, "since to be sure there was no like redness in the water, nor in the leaves, nor in the oil."[6] His reply is as follows—the new color derives from the fact that the penetrating corpuscles of oil of tartar have entered into the leaves, drawing forth very subtle particles from them and thus altering the combined texture of the mixture (*texturam ipsius commutent*). At this point Gassendi makes a move that could have led him to attempt a reduction to the pristine state. In order to emphasize his point that only a change in texture has occurred, Gassendi now repeats the same experiment but substituting oil of vitriol (sulfuric acid) for oil of tartar. The new solution of senna and oil of vitriol does not become red as it did before, but Gassendi points out that if rose leaves are now substituted for the senna, a red color will again emerge. On the other hand, he adds, a solution of rose leaves and oil of tartar does not redden. This experiment, which demonstrates both the selective "power of dissecting, moving, and converting" of different chemicals and the fact that color change can result from alterations in

4. For the claim that the chymical principles are *moleculae*, see Pierre Gassendi, *Animadversiones in decimum librum Diogenis Laertii* (Lyon: Guillaume Barbier, 1649), p. 398. For the claim that salt and sulfur may be responsible for tastes and smells, see Gassendi, *Syntagma philosophicum*, in *Petri Gassendi Dinensis... opera omnia* (Lyon: Laurentius Anisson & Ioannes Baptista Devenet, 1658), vol. 1, pp. 411A–414B.

5. Boyle, *Colours*, in Hunter and Davis, *Works*, vol. 4, pp. 147–150.

6. Gassendi, *Animadversiones*, p. 228: "Unde–nam hoc vero: quippe nullus rubor consimilis fuit neque in aqua, neque in foliis, neque in ipso oleo."

texture, reveals both the proximity and distance between Gassendi on the one hand and Boyle and Sennert on the other. To put matters briefly, it seems simply not to have occurred to Gassendi to employ an acid to restore a basic solution to its original color or vice versa and hence to perform a reduction to the pristine state, despite the fact that he was using acids and bases on the same substances to generate differently colored solutions.[7]

Another case where Gassendi neglected an obvious reduction to the pristine state occurs in his discussion of generation and corruption. In an argument intended to defeat Aristotle's claim that an ingredient present in much smaller volume than another can lose its own form and be transmuted into the one of greater volume when the two are combined, like a drop of wine placed in a vat of water (*De generatione et corruptione* I 10 328a28–31), Gassendi adduces the case of electrum, a gold-silver alloy. As he points out, one ounce of silver can be fused with thousands of gold, so that the silver becomes invisible to the eye. Does it follow, then, that the silver will lose its form and become gold, just as Aristotle's wine became water? "Not at all," replies Gassendi, "in fact it happens that it can be picked out (*seligi*) and recovered by means of aqua fortis and extracted from the mass, as will appear. For not only will whatever silver was in every very tiny particle of the mass be extracted, but it will also be drawn forth from the surface of the whole, so that the whole surface will be left pierced, roughened, and as if eaten out by worms."[8] It is instructive to compare Gassendi's example with the dissolutions and reductions of silver and electrum described by Sennert and Boyle. In Gassendi's case one finds no filtering of the dissolved metal with paper to demonstrate the

7. In fact, Boyle would later use acid spirits to return the reddened senna infusion to its original color, hence indicating the feasibility of converting Gassendi's experiment into a *reductio in pristinum statum*. See Boyle, *Usefulness of Natural Philosophy*, part 2, in Hunter and Davis, *Works*, vol. 3, p. 370. Since Boyle is rather vague on this point, I tried the experiment myself in a laboratory in the Chemistry Department at Indiana University. The senna proved to be an effective color indicator. Infusing a small quantity of dried senna leaves (obtained from a botanical supply house) in warm water for a few minutes with agitation, I received a clear solution colored golden yellow. I filtered this and added dissolved potassium carbonate (oil of tartar), whereupon the solution immediately became red. When a few drops of concentrated sulfuric acid (oil of vitriol) were added to this, the solution returned to its former yellow color. Repeating the process several times produced the same results each time. For Boyle's further comments on Gassendi's experiment, see Boyle, *Colours*, in Hunter and Davis, *Works*, vol. 4, pp. 147–150.

8. Gassendi, *Animadversiones*, p. 401: "Nequaquam sane; & vel ex eo constat, quod potest aqua forti seligi, educique ex massa: veluti patebit, cum non modo ex minutissima quaque particula massae extrahatur quicquid erit in illa argenti, sed ex circumferentia etiam totius ita educetur, ut superficies tota punctata, asperataque, & quasi tineis ita exesa remaneat."

small size of its corpuscles, nor any reference to a particulate precipitate at the bottom of the vessel. Remarkably, Gassendi has not even taken the step of precipitating the silver that has been dissolved by his aqua fortis, which would have led to a reduction to the pristine state. Nor does this omission stem from ignorance on his part, since elsewhere he very clearly describes the precipitation of dissolved metals, though again not in the context of a demonstrative return to the pristine state.[9] For Gassendi's purposes, his dissolution of the silver in a sample of electrum had served quite well. But the experiment was not intended to demonstrate the particulate structure of the silver; its purpose, rather, was to show that the silver had not been transmuted into gold. It is not too much to say that Gassendi, unlike Sennert and Boyle, was simply not concerned with the direct experiential proof of corpuscular microstructure that the *reductio in pristinum statum* seemed to offer. He was not an Aristotelian empiricist of Sennert's stamp, stubbornly refusing to transgress the limits that the senses and laboratory apparatus presented to him, nor was he a Baconian like Boyle, wedded to the approach of the experimental natural history that would lead him to subject his matter to every available test. In short, Gassendi's failure to make full use of the reduction to the pristine state once again throws into high relief the close association between Boyle's approach and that of Sennert.

ESSENTIAL AND EXTRAESSENTIAL ATTRIBUTES

Gassendi, nonetheless, would provide Boyle with important experiments in the latter's attempt to weave together a coherent attack on hylomorphic theories of quality. It has long been recognized that Boyle's attempts to demonstrate the mechanical origin of qualities owed a significant debt to Gassendi's approach, which frequently consisted of attempting to induce or remove qualities by alteration of texture alone.[10] As we will now see, Boyle developed such arguments still further and combined them with Sennert's reductions to the pristine state to produce

9. Gassendi, *Animadversiones*, p. 351: "Quo loco mirabile est, granula non modo argenti, verum etiam auri, cum ponderosiora sint aquae fortis, aut regalis corpusculis, nihilominus in ea sustenari. Sed causa fortassis sunt commisti sales, qui per aquam fusi, & quadam cohasione ab imo usque sese invicem sustenantes, granula metalli a se corrosa, complexaque sustineant, indicioque est, quod ubi guttis aliquot olei tartari praeinfusis communis aqua infunditur, tum metalli granula fundum petiunt, quasi scilicet nova aqua subeunte huiusmodi saleis, illorum cohaesio solvatur, seu, continuatio interrumpatur, sicque contenta granula pondere suo statim subsidant."

10. See, for example, Anneliese Maier, *Die Mechanisierung des Weltbilds im 17. Jahrhundert* (Leipzig: Felix Meiner, 1938), pp. 58–59.

a powerful attack on theories of Aristotelian inspiration linking qualities to form. In the course of making his arguments, however, Boyle would be forced to deal with that most Aristotelian of questions—what is it that truly identifies a thing as itself and distinguishes it from others? What are the essential characteristics of things as opposed to their merely accidental qualities? A key text for Boyle's discussion of this is his "History of Particular Qualities," published in 1670 as an introduction to *Cosmical Qualities*.

In "Particular Qualities," Boyle is concerned inter alia with an objection that an opponent of the mechanical philosophy might pose—if a given quality, say the color white, is due to a particular "mechanical contexture or fabrick," then one might suppose that two different white bodies should agree in their other qualities as well. Why is it, then, to cite one counterexample, that the volatile salt of blood, which is white, has a strong scent while many other white calces do not?[11] Boyle gives several possible answers, but the one that concerns us most begins by observing that it is possible for different types of aggregate corpuscles to take on mechanical affections producing the same sensible quality and for the same type of aggregate corpuscles to take on mechanical affections producing different sensible qualities. We will focus on the latter case, as when iron is heated by hammering, or silver made springy:

oftentimes Corpuscles of very differing natures, if they be but fitted to convene, or to be put together after certaine manners, which yet require no radicall change to be made in their Essential Structures, but only a certaine juxtaposition or peculiar kind of Composition; such Bodies I say may notwithstanding their Essentiall differences exhibit the same Qualitie. For Invisible changes made in the minute and perhaps undiscernable parts of a stable Body may suffice to produce such alterations in its Texture, as may give it new Qualities, and consequently differing from those of other Bodies of the same kind or Denomination, and therefore though there remains as much of the former structure as is necessary to make it retaine its Denomination, yet it may admit of alterations sufficient to produce new Qualities: Thus when a Barr of Iron has been violently hammered, though it continues Iron still, and is not visibly altered in its Texture; yet the Insensible parts may have been put into so vehement an Agitation, as may make the Barr too hot to be held in one's hand. And so if you hammer a long and thin peice of Silver, though the change of Texture will not be visible; it will acquire a springyness that it had not before.[12]

11. Boyle, "History of Particular Qualities," in Hunter and Davis, *Works*, vol. 6, p. 279.
12. Boyle, "History of Particular Qualities," in Hunter and Davis, *Works*, vol. 6, p. 279.

The interesting thing about this passage is Boyle's explicit distinction between the "essential structure" of the corpuscles and their induced characteristics brought about by the motion and altered texture as when a bar of iron is hammered. The iron remains iron, whether hammered or not, as a result of its "essential structure." This "essential structure" refers to the makeup of the aggregate corpuscles that supply the fundamental qualities of iron as such. Boyle continues this idea by considering how bodies of very different quality can be made reflective:

> If on the surface of a Body there arise or be protuberant a multitude of Sharp and stiffe parts, placed thick or close together, let the body be Iron, Silver, or Wood, or of what matter you please, these extant and rigid parts, will suffice to make all these Bodies to exhibit the same Quality of Asperity or Roughnes. And if all the extant parts of a (Physicall) superficies be so depressed to a Level with the rest, that there is a coaequation, if I may so speak, made of all the superficiall parts of a Body; this is sufficient to deprive it of former Roughnes, and give it that contrary Quality we call Smoothnes. And if this Smoothnes be considerably exquisite, and happen to the Surface of an Opacous Body of a close and solid contexture, and fit to reflect the incident Rays of Light and other Bodies unperturb'd, this is enough to make it specular, whether the Body be Steele, or Silver, or Brasse, or Marble, or Flint, or Quicksilver, &c."[13]

In this passage, Boyle presents the inverse of his iron example. Just as iron remains iron whether its particles are set into rapid internal motion by a hammer or not, so different materials, such as steel, brass, or marble, can all acquire the same smooth surface texture that leads to reflectivity. Again the "essential structure" of the different materials remains unchanged, but the "sharp and stiff parts" that protrude from their bodies are uniformly depressed to present a reflective surface. Boyle goes on to say that a similar change in the inessential attributes of matter can occur in the production of sounds. It makes little difference whether a particular pitch is produced by gut strings or by a metallic bell: if the "waving motions" into which the air is set by the string or bell are the same, the sound that is produced will be the very same note.[14] In the case of the string and bell not only will the material of the musical instrument differ, but the structure of the instrument will be vastly different—what remains the same in both instances is the texture of the air and the motion into which it is put. But in all the cases that we have considered, the corpuscles that act as the vector of the quality in question—be they

13. Boyle, "History of Particular Qualities," in Hunter and Davis, *Works*, vol. 6, p. 280.
14. Boyle, "History of Particular Qualities," in Hunter and Davis, *Works*, vol. 6, p. 280.

composed of iron, wood, air, or other material—remain unchanged. The particles are being set into new relationships to one another, thus creating new and altered textures, but they themselves might as well be Sennert's "minima of their own genus." In order to make this point even more clearly, Boyle adds the following observations:

> For here it is to be considered, that besides that peculiar and Essentiall Modification which constitutes a Body, and distinguishes it from all other that are not of the same Species, there may be certain other Attributes that we call *Extra-essentiall;* which may be common to that Body with many others, and upon which may depend those more externall Affections of the Matter which may suffice to give it this or that Relation to other Bodies, divers of which Relations we stile Qualities. Of this I shall give you an Evident Example in the Production of Heat. For provided there be a sufficient and confus'd Agitation made in the insensible parts of a Body, whether it be Iron or Brasse, or Silver, or Wood, or Stone, that vehement Agitation without destroying the Nature of the Body that admits it, will fit it for such an Operation upon our sense of Feeling, and upon Bodies easy to be melted (as Butter, Wax &c.) as we call Heat.[15]

Having argued that bodies of the same material can take on divergent qualities and that bodies of vastly different material can produce the same qualities, Boyle now introduces a new term for these characteristics that do not depend on the essential structure of the corpuscles— "extraessential attributes." His idea here maps loosely onto the modern distinction between chemical and superficial physical change. The heat that one feels upon rubbing iron, brass, silver, wood, or stone can be induced and allowed to depart without altering the chemical properties of the material being rubbed, which are taken as the measure of its essential character. The same may be said of the polishing that produces shininess in wood or audible vibration in air: both are physical states induced without affecting the chemical structure of the molecules involved. From Boyle's perspective, the distinction between "essential" and "extraessential" attributes is a clear articulation of the difference between those qualities that necessarily follow the structure of a particular aggregate corpuscle and those that can be imposed or deleted mechanically without altering its essence. But here an interesting problem emerges—how did Boyle know which properties of a body were essential if in fact an indeterminate number of its qualities—both sensible and insensible—could be

15. Boyle, "History of Particular Qualities," in Hunter and Davis, *Works*, vol. 6, p. 280.

modified mechanically? Here again we see the necessary connection between Boyle's chymical experimentation and his mechanical philosophy, for it was above all the classification into chymical species that allowed Boyle to determine the essential differences of the aggregate corpuscles. *Colours* contains large sections devoted to indicator tests for deciding whether a particular substance belongs to the class of "acid salts," "alkalizate salts," or "urinous salts."[16] In other contexts, he employs such time-honored tests as cupellation, dissolution in different mineral acids, and color of flame to detect a metal or other substance when its presence is not obvious to the senses.[17]

Many of these tests are based on the assumption that the aggregate corpuscles being tested for are not destroyed by the test itself—instead they remain undivided during the procedure and hence retain their identity. At the same time, the reagent employed to reveal the hidden substance is assumed to act selectively on the latter's aggregate corpuscles (by causing them mutually to disperse or coalesce, for example) and hence to circumvent the sort of generalized mechanical effect that Boyle describes as "extraessential." Chymical indicator tests therefore allowed Boyle a way to avoid the deceptive realm of extraessential properties and to arrive at genuine species-determining characteristics. One can see once again that Boyle's mechanical philosophy and chymistry were deeply integrated: it was chymistry that allowed him to distinguish the essential differences of bodies in a relatively certain fashion, and without such stable essences Boyle could not argue that the qualitative mutability of the phenomenal world was mostly a matter of alterations in texture imposed on fundamentally unchanged corpuscles by mechanical means.

BOYLE'S EXPERIMENTAL STRATEGIES AND THE INDUCTION OF EXTRAESSENTIAL PROPERTIES

It may at first evoke surprise that one should employ the Aristotelian language of "essential" characteristics in the mechanical philosophy. Boyle himself was sensitive to this issue, which led him explicitly to consider

16. Boyle, *Colours*, in Hunter and Davis, *Works*, vol. 4, pp. 106–109, 125, 154–157. For Boyle's tests designed to reveal these different types of salt, see Newman and Principe, *Alchemy Tried in the Fire*, pp. 275–281.

17. I do not mean to suggest that these tests were the only means for determining the identity of substances at Boyle's disposal—in certain circumstances, specific weight or other physical properties would have served as well.

the scholastic distinction between two ways in which the term "accident" could be considered. Boyle's reading of this distinction is as follows. In the sense of *accidens praedicabile* (predicable accident), an accident is distinguished from an essential attribute. When one says that a wall is white, the speaker is predicating whiteness of the wall. But since a wall can be either white or not white, the predicable accident "whiteness" is not essential to the wall as a wall.[18] In the sense of *accidens praedicamentale* (predicamental accident), on the other hand, accident is opposed to substance—to be an accident means to be something that cannot subsist of itself but must be a characteristic of something else. Returning to Boyle's example of the white wall, the wall's whiteness as a "predicamental" accident cannot exist without the wall, while the wall itself is able very well to exist without whiteness.[19]

In his own treatment of qualities, Boyle employs the distinction between these two senses of accident in these words—"Nor need we think that Qualities being but Accidents, they cannot be *essential* to a Natural Body; for Accident, as I formerly noted, is sometimes oppos'd to Substance, and sometimes to Essence: and though an accident can be but accidental to Matter, as it is a Substantial thing, yet it may be essential to this or that particular Body."[20] In other words, a particular quality may be accidental to the matter in which it happens to be found in the predicamental sense and yet in the predicable sense it may actually belong to the particular essence of a body made out of that matter. To clarify this, Boyle uses an example from Aristotle (*Metaphysics* 1049b4–1051a2). In the case of a brass sphere, the roundness is (predicamentally) accidental to the matter, since the matter can exist without roundness but the roundness cannot exist without matter. Nonetheless, roundness

---

18. "Whiteness" is not part of the definition of "wall," so it cannot be said to belong essentially to the species of walls.

19. Boyle, *Forms and Qualities*, in Hunter and Davis, *Works*, vol. 5, p. 308. The term *praedicamentale* comes from Aristotle's ten categories, the *praedicamenta* (substance, quantity, quality, relation, place, time, situation, state, action, and passion). For more on the distinction between *accidens praedicabile* and *accidens praedicamentale*, see Rudolph Goclenius, *Lexicon philosophicum* (Frankfurt: Petrus Musculus, 1613; Olms reprint, 1964), pp. 26–32. As Goclenius points out, the fact that something is one kind of accident does not necessarily mean that it is also the other. When I say that a table is wooden, the wood of the table is not a predicamental accident, since it can subsist of itself, but it is a predicable accident of the table, since not all tables are made of wood. If I say that white is a color, color is not a predicable accident, since it is the genus into which whiteness falls; color is a predicamental accident, though, since it cannot subsist of itself.

20. Boyle, *Forms and Qualities*, in Hunter and Davis, *Works*, vol. 5, p. 324.

is essential to the brass sphere when considered not just in terms of its matter but also in terms of itself as a brass sphere. In the same way, Boyle wants to treat essence as a "convention," that is, a collection, of qualities pertaining to a particular body. Insofar as one is considering a body with particular qualities, one can speak of those qualities as providing an essential definition for the body, even if the qualities are not essential to the uniform, catholic matter out of which the body is ultimately composed. There is no underlying ontology of substantial forms implied in this conception of essences. Yet Boyle adds that even the scholastic term "form" is permissible in this context, so long as it is acknowledged to pertain to the aggregation of accidents in a body rather than signifying a self-subsistent substance distinct from matter. In short, Boyle's analysis of *accidens praedicabile* and *accidens praedicamentale* allows him to use the idea of "essence" in his natural philosophy without committing him to the existence of substantial forms, which it is his primary goal to destroy.

When we descend from these verbal distinctions to practice, it becomes evident that the effort Boyle put into determining what were effectively the specific differences and essences of various materials was not an anomaly, nor was it testimony to the divergence of his "science" (in Alan Chalmers's sense) from his mechanical philosophy. To the contrary, it was fundamental to his quest to demonstrate the mechanical origin of qualities. The essential attributes of the aggregate corpuscles allowed Boyle to link these bits of matter to the experiences of the laboratory. How exactly did this work? Many scholars have stressed that Boyle's corpuscular philosophy was experimental in character. But the precise character of Boyle's experimental program and its relationship to his mechanical philosophy has seldom received the detailed treatment that it deserves, leading to the problematic claims discussed in the previous chapter. In the following part, I will focus on three experimental strategies used by Boyle that rely implicitly on his appeal to the essential attributes emerging from aggregate corpuscles. For the validation of Boyle's mechanical philosophy, corpuscles endowed with "chymical qualities" were of fundamental importance.

The casual observer could easily assume that the goal of Boyle's many experimental treatments of the mechanical philosophy was to prove by laboratory means that the material world is composed of physically indivisible atoms endowed only with the "catholic affections" or primary qualities of size, shape, and motion. But this is not really the case. Instead of demonstrations of this sort, one does find him employing occasional experiments intended to reveal the semipermanence of the "primary clusters" or aggregate corpuscles, usually employing the reduction to the pristine state.[21] We have already discussed the most conspicuous of these demonstrations, namely, the one occurring at the beginning of part 1 of *The Sceptical Chymist*, where Boyle points out that gold can be dissolved in aqua regia, filtered through paper, and then reduced to "the self-same Numerical, Yellow, Fixt, Ponderous and Malleable gold it was before its commixture."[22] Boyle's highlighting of the experiment stemmed from the fact that it supplied the two observational criteria necessary for the verification of his theory of "primary clusters." First, the disappearance of the gold and its passage through filter paper ensured that it had been broken down into very small particles by the acid. Second, the fact that the gold could be regained intact showed that its particles had not been divided into *prima naturalia* by the aqua regia, despite the visible dissolution of the gold in the acid. Since aqua regia was the only "vulgar menstruum" known to attack gold, the resistance of its "primary clusters" to this strong corrosive was powerful evidence of their semipermanent character.

Yet as Boyle himself said many times, the main goal of his experimental program was not the mere demonstration that there is a corpuscular microworld or the fact that there are such things as semipermanent corpuscles, but the reduction of "particular qualities"—the phenomena of the sensible world—to mechanical causes. Boyle's general experimental procedure in this project was threefold, as the following passage from his *Experiments, Notes, &c. about the Mechanical Origine or Production of Divers Particular Qualities* (1675—henceforth *Mechanical Qualities*) indicates.

> I shall adde on this occasion, that there are three distinct sorts of Experiments (besides other proofs) that may be reasonably employ'd, (though they be not equally efficacious) when we treat of the *Origine* of Qualities. For some Instances may be brought to shew, that the propos'd Quality may be Mechanically *introduc'd* into a portion of matter, where it was not before. Other Instances there may be to shew, that by the same means the Quality may be notably

21. Straightforward references to the *reductio ad pristinum statum* may be found in Boyle, "Excellency of the Mechanical Hypothesis," in Hunter and Davis, *Works*, vol. 8, p.113 (mercury and gold); *The Sceptical Chymist*, in Hunter and Davis, *Works*, vol. 2, pp. 230–231 (gold and mercury); *Forms and Qualities*, in Hunter and Davis, *Works*, vol. 5, p. s326 (mercury, camphor), pp. 395–398 (camphor).
22. Boyle, *The Sceptical Chymist*, in Hunter and Davis, *Works*, vol. 2, p. 230.

varied as to degrees, or other not essential Attributes. And by some Instances also it may appear, that the Quality is Mechanically *expell'd* from, or *abolish'd* in, a portion of matter that was endow'd with it before. Sometimes also by the same Operation the former quality is destroyed, and a new one is produc'd.[23]

The induction, elimination, and alteration of secondary qualities by mechanical means allowed Boyle to argue in general terms that such qualities were at basis mechanical. Experiments of this sort did not require that one make any assumption about the microstructure of a particular substance except that it was composed of semipermanent corpuscles that could be rearranged or altered in their motion to yield a new texture. The premise of semipermanent corpuscles, in turn, had acquired experimental justification from the reduction to the pristine state, as in the reduction of gold from its compounds in *The Sceptical Chymist*. An example of the mechanical removal and replacement of secondary qualities appears also in *The Sceptical Chymist*, where Boyle, in the guise of the interlocutor Carneades, alters the sensible qualities of lead without removing or adding any material components:

> And to let you see, *Eleutherius*, that 'tis sometimes at least, upon the Texture of the small parts of a body, and not alwaies upon the presence, or recesse, or increase, or Decrement of any one of its Principle, that it may lose some such Qualities, and acquire some such others as are thought very strongly inherent to the bodies they Reside in; I will add to what may from my past discourse be refer'd to this purpose, this Notable Example, from my Own experience; That Lead may without any additament, and only by various applications of the Fire, lose its colour, and acquire sometimes a gray, sometimes a yellowish, sometimes a red, sometimes an *amethistine* colour; and after having past through these, and perhaps divers others, again recover its leaden colour, and be made a bright body.[24]

Here we see the "small parts" or aggregate corpuscles bearing the essential properties of lead being subjected to color changes by fire, which

---

23. Boyle, *Mechanical Qualities*, in Hunter and Davis, vol. 8, p. 322.
24. Boyle, *The Sceptical Chymist*, in Hunter and Davis, vol. 2, p. 342. Boyle's description of lead was probably inspired by Gassendi. See the following passage from Gassendi's *Syntagma philosophicum*, in *Petri Gassendi Dinensis . . . opera omnia*, vol. 1, p. 380A: "Quo loco mirabile est, quod Chymici norunt, plumbum, cum res adeo densa, & opaca, sit, ubi urgetur igne vehementissimo, intereaque insufflatur, abire, contexi, conformari in quandam quasi hyacinthi speciem, quae sit perspicua admodum; tantum ad perspicuitatem valet non raritas modo, verum etiam partium certa dispositio. Omitto autem, ut igne rursus suam plumbi speciem repetat, inversa rursus dispositione partium."

Boyle assumes to be a mechanical agency.[25] As in the reduction to the pristine state of gold dissolved in aqua regia, the initial lead bearing its original properties is recovered at the end of the experiment. This recovery ensures that no substantial change has occurred—the lead has persisted beneath the various disguises that it has assumed. In the text of *The Sceptical Chymist* Boyle follows this with two further experiments— the lead can be "made as brittle as glass" by heating alone and then returned to its former malleability. Similarly, heating lead can transform it into a transparent substance, which can again be returned to its former opacity by altering "the manner and method of exposing it to the fire." In each case the lead is returned to its pristine state for the same demonstrative purpose: Boyle wants to show that the qualities imposed by heating are, as he put it in "Particular Qualities," "extraessential." In order to make the claim that these mechanically induced properties were not essential to but emergent from matter, however, Boyle obviously required a material substrate from which they could emerge. And since the point of his argument was that the material substrate remained essentially unchanged, he required semipermanent corpuscles whose rearrangement, shifting, and other local change accounted for the changes in color. He could not invoke the primary qualities of the *prima naturalia* here, for he had no experimental verification that such minimal corpuscles, unlike his aggregate corpuscles, existed. Hence the empirical force of Boyle's claim that mechanical change at the macrolevel induces mechanical change at the microlevel, resulting in a change of sensible quality, relied on the demonstrated existence of semipermanent aggregate corpuscles that persisted through the induction and removal of extraessential qualities.

Boyle's general claim that mechanical change at the observational level produced an "access, recess, or transposition" of aggregate corpuscles required two prerequisites—that the agencies of change employed be mechanical and that the insensible corpuscles really exist. Following the lead of Francis Bacon's kinetic theory of heat, Boyle claimed that the heat employed to alter his lead was a mechanical phenomenon. As for the insensible corpuscles, he had demonstrated the existence of Sennertian "minima of their own genus" by means of reductions to the pristine state. But how could he progress to the further claim that

25. Heating was a purely mechanical operation to Boyle, on grounds that he inherited from Francis Bacon. See Boyle, "Of the Mechanical Origin of Heat and Cold," in Hunter and Davis, *Works*, vol. 8, pp. 343–344.

any particular sensible quality was the result of a particular mechanical attribute at the microlevel? Boyle's famous reticence to commit himself to specific mechanical models reveals his cognizance of this dilemma.[26] He was aware of the fundamental problem inherent in transdiction, the passage from observational entities to those that are inherently nonobservational: how could we acquire empirical verification of that which must lie outside our experience? Maurice Mandelbaum, in his famous treatment of this subject, identified two strategies by which Boyle passed from the sensible world to the insensible. The first involved "the principle of extending our sense knowledge through analogy," the second a principle that Mandelbaum called "the translation of explanatory principles from the observed to the unobserved."[27] As an example of transdiction by analogy, Mandelbaum invokes a passage from Boyle's "History of Fluidity" (1661), where compressed snow is used to demonstrate the increased hardness that comes with reduction of porosity: this principle is then contingently assumed to obtain at the microlevel as well. For his translation principle, Mandelbaum refers to a famous section of "The Excellency and Grounds of the Corpuscular or Mechanical Philosophy" (1674), where Boyle argues that laws of nature such as the acceleration of falling bodies and "the laws of motion" more generally apply both to large bodies and small: there is no reason to think that they should cease to apply at the microlevel.[28]

Mandelbaum's translation principle is similar to yet another strategy employed by Boyle, where appeal is made to decreasing stages of size. I will call this additional strategy for acquiring knowledge of the microworld "transdiction by substantial identity." A clear example of this move may be seen in another passage from the "History of Fluidity." Boyle begins this argument by postulating that fluidity at the macrolevel is largely due to the corpuscles of a substance being extremely small. In order to illustrate this principle, Boyle begins with a straightforward analogy. Let us consider a number of bags filled with different types of objects. Let one bag contain apples, another walnuts, another filberts,

---

26. See the discussion of this Boylean feature in Sargent, *Diffident Naturalist*, pp. 105–106, 125. The programmatic character of Boyle's corpuscular claims is also alluded to briefly by E. J. Dijksterhuis in his *The Mechanization of the World Picture* (Oxford: Clarendon Press, 1961), p. 437, as Sargent notes.

27. Maurice Mandelbaum, *Philosophy, Science, and Sense Perception* (Baltimore: Johns Hopkins Press, 1964; paperback edition, 1966), 110–111. Mandelbaum adds a third principle, "the method of indirect confirmation that all scientists must employ" (p. 112). Since this principle is more general than transdiction, however, it is not relevant to the present discussion.

28. Mandelbaum, *Philosophy, Science, and Sense Perception*, 109–110.

yet another wheat, a fifth one sand, and a sixth one flour. If we then pour out the contents of each bag, we shall see that those containing finer components resemble more closely the action of a fluid.[29] Now Boyle could have left this as an analogy, as he does with his many comparisons between microlevel processes and those of clocks, balances, water mills, and other machines.[30] If he had relied on the analogy of sacks filled with different kinds of items, Boyle would have had only an illustration of a possible situation at the insensible level, however. Instead, he carried his investigation one step further. Several pages after the analogy, Boyle proceeds to a discussion of niter, which relies on the reduction of its particle size by mechanical means—

> And hence we may proceed to consider, what Fluidity Salt-Petre is capable of without the intercurrence of a Liquor: and this may be two-fold. For first, if it be beaten into an impalpable powder, this powder, when it is pour'd out, will emulate a Liquor, by reason that the smallness and incoherence of the parts do both make them easie to be put into motion, and make the pores they intercept so small, that they seem not at a distance to interrupt the unity or continuity of the Mass or Body. But this is but an imperfect Fluidity, both because the little grains or Corpuscles of Salt, though easily enough moveable, are not alwaies in actual motion; and because they continue yet so big, that both they and the spaces intercepted betwixt them are, near at hand, perceivable by sense. But if with a strong fire you melt this powder'd Nitre, then each of the saline Corpuscles being sub-divided into I know not how many others, and these insensible parts being variously agitated by the same heat, (both which may appear by their oftentimes piercing the Crucible after fusion, wherein they lay very quietly before it) the whole body will appear a perfect Liquor . . . and such also is the Fluidity of melted metals.[31]

29. Boyle, "History of Fluidity," in Hunter and Davis, *Works*, vol. 2, p. 127. The passage is probably inspired by Gassendi. See Gassendi, *Animadversiones*, pp. 333–334: "Neque vero obstat, quod aqua videatur esse quidpiam continuum, contra quam acervus frumenti videtur; id enim oritur ex eo, quod quo grana minutiora sunt, eo spatiola intercipiantur, quae insensibiliora sint, corpusque ex ipsis constans videatur minus interruptum, seu, quod est idem, magis continuum; prout licebit intelligi, si cum acervo lapidum, contuleris acervum nucum; cum hoc, acervum tritici; cum hoc, acervum sabuli; cum hoc acervum cineris; atque ita de reliquis."
30. See Boyle, "Free Considerations about Subordinate Forms," in Hunter and Davis, *Works*, pp. 458–464.
31. Boyle, "History of Fluidity," in Hunter and Davis, *Works*, vol.2, pp. 131–132. Very likely Boyle's direct inspiration for this passage lies in a parallel one in Gassendi, where metallic powders are employed instead of niter. See Gassendi, *Animadversiones*, p. 334: "At quia granula haec impalpabilia compositissima adhuc sunt, utpote non extreme resoluta in particulas, ex quibus metallum contexitur; fit, ut si praeterea eliquaveris, hoc est, ignem admoveris, cuius corpuscula subeant, & discutiant ista granula; (quod fortis aquae corpuscula, aut limae subtilissimae denticuli nusquam potuerint) fit, inquam, ut metallum, eadem ratione, qua aqua fluat."

In this example, niter is observed first in the form of a fine powder, which, like the grains of wheat, sand, and flour in his earlier analogy, acquires some approximation to fluidity. If one looks at the niter from a distance as it is being poured out of its vessel, the pores between the particles of niter cannot be seen, and the flowing salt acquires the illusion of continuity. Yet the fluidity is not complete, since the particles are still too big to be set into motion by the ambient heat. This condition can be corrected, however, by supplying more heat, with the result that the particles are divided further and set into motion. If one accepts Boyle's premise that heat is a purely mechanical phenomenon that sets particles into motion and thus drives them apart, the niter experiment acquires considerable cogency. By mere "grinding" of the same substance into corpuscles at different levels of minuteness, one proceeds from a solid to a liquid state—no other variable has been introduced. Here we pass from analogy at different levels—as in the case of the bags filled with discrete items—to identity at different levels. For this reason, I refer to the niter experiment as an example of transdiction by substantial identity. The transposition of the same aggregate corpuscles into different states allows Boyle to go from the observational world to the microlevel without relying on mere analogy.

In the first of the examples mentioned above, where Boyle imposed a wide array of phenomenal changes on lead and then reduced it to its pristine state, the substantial identity of the lead throughout the changes (demonstrated by its subsequent reduction) served as a guarantor that the alterations in quality were the result of mere changes in texture. The final Boylean strategy that I wish to discuss relates to what could be seen as the inverse procedure—where the same effect is induced in a wide variety of substances, as when the same color is produced by crushing or grinding different materials. Boyle describes this Baconian demonstration in some detail in "Particular Qualities," where he addresses the objection that "if two bodies agree in one quality, and so in the structure on which that quality depends, they ought to agree in other qualities also."

And so in the Instance nam'd in the Objection about Whitenes. T'is accidentall to that Quality that the Corpuscles it proceeds from should be little Hemispheres [i.e.,bubbles]. For though it happen to be so in Water agitated into froth; yet in water frozen to Ice, and beaten very small, the Corpuscles may be of all manner of Shapes; and yet the powder be white. And it being sufficient to the produceing of Whitenes that the incident Light be reflected copiously every way and untroubled by the reflecting Body, it matters not whether that

Body be Water, or white Wine, or some other clear Liquor turn'd into froth, or Ice, or Glasse, or Christall, or Clarified Rozin, &c. beaten into Powder; since without dissolving the Essentiall Texture of these formerly diaphanous Bodies, it suffices that there be a comminution into graines numerous and small enough by the multitude of their surfaces, and those of the Aire (or other fluid) that gets between them, to hinder the passage of the beames of Light, and reflect them every way as well copiously, as vnperturb'd.[32]

Here Boyle begins by invoking the whiteness produced by the mere "comminution" of various substances, such as froth, ice, glass, crystal, and resin. Clearly the demonstration presupposes the principle of substantial identity as a precondition for induction of extraessential qualities, since each material is independently ground or agitated to make it white. The "essential structure" provided by the corpuscles remains unaltered, while its mechanical treatment causes it to take on the new extraessential quality of whiteness. At the same time, the production of the single new effect, whiteness, depends in every case on "comminution into grains numerous and small enough" to reflect the incident light in all directions. Hence, as in the case of transdiction by substantial identity, we are passing from the large to the small, though not necessarily to the insensibly small. It is possible to see, then, that the strategy of producing an identical effect by a similar mechanical modification of different substances can contain elements of the former two strategies. In effect, this third strategy supplies a means of testing the results obtained by the former two approaches, and perhaps others, over a broad sample. It is part of Boyle's larger experimental approach inherited from Francis Bacon.[33] And yet, insofar as this third strategy assumes a lack of substantial alteration on the part of its subject, it too relies on Boyle's empirical demonstration of the reality of semipermanent aggregate corpuscles derived from Daniel Sennert.

To summarize, then, we have shown that Boyle relied on Sennert's reductions to the pristine state in two quite different, though related, ways. First, the Sennertian examples of metals simply dissolved in acids and then reduced or metals combined with other substances and then restored to their former state by reduction, provided Boyle with the necessary evidence to make the claim that "minima of their own genus"

32. Boyle, "The History of Particular Qualities," in Hunter and Davis, *Works*, vol. 6, pp. 280–281.
33. For Bacon's own experiments with crushed glass and froth made from agitated water, see aphorism 23 of *Novum Organum*, part 2, in Spedding et al., *The Works of Francis Bacon* (London: Longman, et al., 1860), vol. 4, pp. 156–158. See Sargent, *Diffident Naturalist*, pp. 22–61, for more on Boyle's debt to Bacon in this area.

really did exist in nature. He even employs Sennert's use of filter paper to reveal the minute size of these corpuscles in the highly prominent second proposition of *The Sceptical Chymist*, part 1. But this is only the beginning for Boyle. In addition, he extends the reduction to the pristine state to form an integral part of his program of revealing the mechanical origins of "particular qualities." The appeal of the reduction to the pristine state here lies in its role as a guarantor that no substantial change has occurred in the corpuscles undergoing extra-extraessential modification by mechanical means. Boyle assumes, just as Sennert had done, that these aggregate corpuscles do not lose their essential nature at the beginning of a process only to regain it unchanged at the end. If Boyle's composite corpuscles are themselves experiencing no essential modification, then the phenomenal change that they undergo must result from an alteration in their mutual texture, brought on by association, dissociation, or transposition. Like Sennert's "one form [that] can go forth into the theater of nature under diverse bodily shape," the essence particular to the aggregate corpuscle remains unchanged while the qualities stemming from its texture (in combination with other corpuscles) undergo alteration if the texture changes.[34] Boyle then applies these results to his strategies of transdiction by substantial identity and his induction of the same qualitative change in a wide variety of materials. Although we have already remarked on Boyle's debt to Sennert for the experimental foundations behind these approaches, however, it is clear that the German academic, unlike his British heir, had no desire to reduce all phenomenal change to the realm of the mechanical. To the contrary, Sennert thought that locking the substantial forms within their corpuscular shells would provide a role for them in a world where the analytical techniques of the laboratory were rapidly rendering hylomorphism incapable of explaining chymical change. For Sennert, substantial form could still provide the efficient causality behind chymical interactions while also supplying the principle of substantial identity to different materials.

REDINTEGRATION AND THE SUBSTANTIAL FORM

We could continue down the path of similarity by pointing out that despite Sennert's grounding in scholastic Aristotelianism and Boyle's

---

34. Sennert, *De chymicorum* (1619), p. 336: "Cum multis, quae in natura fiunt, probabile reddatur, unam formam sub diverso schemate corporis in naturae theatrum progredi posse: Aut certe in uno semine posse esse plures formas, sed subordinatas, ita tamen, ut una sit princeps & Domina, reliquae quasi ministrae."

self-styled allegiance to the "new philosophy," both men profess ne-science when it comes to making categorical statements about the precise origin of the qualities that they deem to be essential. Sennert explicitly adheres to the two principles that substantial form is unknowable and that we only know bodies through their accidents. Boyle, similarly, is notorious for his reticence in making hard claims about the particular mechanical structure behind a given quality. But the fact remains that Boyle's "mechanical hypothesis" postulated the lower level of the *prima naturalia*, which were characterized by the catholic affections alone. Although he was unable to prove the existence of these corpuscles, he looked towards that proof as a goal. Most of the experimental strategies that we have so far discussed implicitly employ the principle of parsimony and burden of proof to make Boyle's larger points. As he looked at it, if he could show that the qualities that we know either directly by sense or by their actions can be induced or destroyed mechanically, why should we make the further assumption that things are different at a lower level of composition? It would be multiplying species beyond necessity to argue otherwise, unless one had additional reasons to do so. Hence it is up to our opponent to supply reasons for thinking that nature, at the most fundamental level, is nonmechanical. In addition, Boyle views his nescience as being of a different order than Sennert's, and he employs the fact that forms are *ex confesso* unknowable to showcase their unintelligibility. How can a hylomorphic opponent expect Boyle to adopt an admittedly unintelligible principle when the operations of machines are perfectly clear and when the manufacture of fine machinery shows that they are capable of translation to the world of the vanishingly small?

Despite the effectiveness of these strategies, Boyle still hungered for a more direct demonstration of the fact that the forms and qualities attributed to bodies really originated from the mere texture of their particles. Unlike the tactics employed so far, this demonstration would attack the putative substantial form of a body directly rather than altering, removing, or inducing accidental qualities while the essential nature of the body remained intact. And yet we will show that here, too, the Sennertian reduction to the pristine state played a part in Boyle's demonstration. A body itself with its own "essential structure," what the Aristotelians would call a substance, would be taken apart into its components and then reassembled, just as the springs and wheels of a watch can be disjoined and shown to have their own properties, such as yellowness and springiness, and then put back together, so that they function together as a watch. In such a case, Boyle reasoned from the principle of parsimony, no one would argue that the "form" of the reassembled body was anything other than the sum of its parts, working in a coordinated fashion. The qualities of the body could no longer be said to flow from the form, since there was no reason to assume that the form itself was anything but a particular structure resulting from the association of its parts.[35]

This reliance on analysis and synthesis is the approach that Boyle pioneered in his early "Essay of Salt-Petre," published in 1661 as part of *Certain Physiological Essays*. He would go on to develop this strategy in *Forms and Qualities* and its 1667 appendix, "Free Considerations about Subordinate Formes"; the appendix is largely a detailed critique of Sennert's theory of subordinate forms. The experiment to which we must now turn is Boyle's famous "redintegration" of niter, a phenomenon that was first discovered by the chymist Johann Rudolph Glauber and probably transmitted to Boyle in the early to mid-1650s by the surgeon and amateur chymist Benjamin Worsley.[36] In simplest terms, Boyle's experiment worked by injecting burning charcoal into molten saltpeter, igniting it. This resulted in the release of nitrogen and carbon in combination with oxygen, leaving a nonvolatile residue of "fixed niter" that resembled salt of tartar or potassium carbonate (in reality it *was* primarily potassium carbonate). Knowing that spirit of niter (nitric acid) could be produced by the thermal decomposition of niter, Boyle then added spirit of niter to the tartarlike residue and acquired a product that resembled the original saltpeter in all its significant properties. Hence he concluded that niter itself is merely a compound of two very different materials, namely, spirit of niter and fixed niter, which we would call an acid and a base.[37] It is important to note that Boyle did not try to produce the spirit of niter from the sample of saltpeter that he was analyzing into its components. Nor did he take full account of the fact that

---

35. In addition to its reliance on the principle of parsimony, Boyle's approach assumes that the "factitious" character of such manufactured substances would rule out the possibility that they had substantial forms on the assumption that only natural things can have substantial forms. This assumption was rendered problematic, however, by the elasticity of the artificial-natural dichotomy. For a discussion of this weakness in Boyle's argument, see Newman, *Promethean Ambitions*, pp. 271–283.

36. See Newman and Principe, *Alchemy Tried in the Fire*, pp. 236–256, for Worsley. See also John T. Young, *Faith, Medical Alchemy and Natural Philosophy: Johann Moriaen, Reformed Intelligencer and the Hartlib Circle* (Brookfield, VT: Ashgate, 1998), pp. 183–216, esp. 198–200.

37. The experiment is clearly described by Boyle, *Certain Physiological Essays*, in Hunter and Davis, *Works*, vol. 2, pp. 92–96.

the charcoal contributed materially to the reaction rather then merely supplying heat. For these and other reasons, he was not able to arrive at what he calls an "equiponderant" analysis and synthesis, where the initial and final products were composed of ingredients that were numerically the same and had the same weight. Although this "adequate" or "equiponderant" redintegration remained a desideratum beyond his abilities, Boyle was keen to show the advantages that such an operation would have in debunking the theory of substantial forms:

> And if upon further and exacter tryal it appears that the whole body of the Salt-Petre, after it's having been sever'd into very differing parts by distillation, may be adequately re-united into Salt-Petre equiponderant to it's first self; this Experiment will afford us a noble and (for ought we have hitherto met with) single instance to make it probable that that which is commonly called the Form of a Concrete, which gives it it's being and denomination, and from whence all it's qualities are in the vulgar Philosophy, by I know not what inexplicable wayes, supposed to flow, may be in some bodies but a Modification of the matter they consist of, whose parts by being so and so disposed in relation to each other, constitute such a determinate kind of body, endowed with such and such properties.[38]

The neatness of this experiment, and the novel use to which Boyle put its results, should not blind us to the sources from which he is deriving the basic structure of his argument. The fact that Boyle sees an equiponderant analysis and synthesis as a desirable goal points to the influence of J. B. Van Helmont, who claimed to have performed similar gravimetric analyses and syntheses of glass and other substances.[39] But the redintegration of saltpeter can also be seen as a Sennertian reduction to the pristine state, even though there is a temporary loss of substantial identity on the part of the niter. Sennert, as we know, viewed dissolution followed by reduction as a *diakrisis* and then a *synkrisis*. He had carefully pointed out in his *De chymicorum* that *diakrisis* can mean either the separation of homogeneous corpuscles from one another, as in the case of silver dissolved in nitric acid, or it can refer to the dissolution of a material into its heterogeneous parts, as when milk separates into butter, whey, and curds. These latter components cannot be reassembled into milk, but that was irrelevant to Sennert, since his main goal did not require this

38. Boyle, *Certain Physiological Essays*, in Hunter and Davis, *Works*, vol. 2, pp. 107–8.
39. See Newman and Principe, *Alchemy Tried in the Fire*, pp. 35–91, where the Helmontian roots of Boyle's gravimetric approach are described at length.

sort of resynthesis. He was intent on disproving the prevailing scholastic theories of mixture by demonstrating the persistence of an underlying substance in the course of phenomenal change, which could be carried out without resynthesizing a compound. It is easy, nonetheless, to see how Boyle could approach the problem from the other end and turn Sennert's proof on its head. Niter was a naturally occurring substance and ought *ex hypothesi* to have a substantial form. But it could be broken into two very distinct materials—each with its own set of properties— and then (in principle) reassembled just as one assembles the parts of a watch. The very arguments that told against the return of a form from its privation, which had worked in favor of Sennert's persistence of the substantial form, now worked against the idea that niter had a substantial form of its own. Erastus and his followers had argued at great length that the loss of a form followed by its immediate reacquisition was impossible within the prevailing hylomorphic theory. If one followed the scholastic maxim that "there is no immediate return from privation to a habit," the properties of the reconstituted niter must result from the fact that its ingredients are "disposed in relation to each other" to form a particular texture.

But again, one must recognize that the equiponderant redintegration of bodies remained an unrealized goal for Boyle. This is borne out by his discussion of the topic in *Forms and Qualities*. Directly before the part of *Forms and Qualities* entitled "Experiments, and Thoughts, about the Production and Reproduction of Forms," Boyle says that he has three kinds of experiments that are particularly suited to rendering his views on the origin of forms probable. The first of these consists of compounding ingredients to produce a body with strikingly different characteristics from those components, the second is redintegration, and the third the transmutation of metals and other materials. When Boyle returns to redintegration later in the text, he observes that it does not merely involve "the first Production" but "the Reproduction of a Physical Body," and that this makes it more qualified to demonstrate the specious nature of the substantial form than the simple compounding of ingredients.[40] But Boyle then proceeds to criticize the optimism of his "Essay of Salt-Petre," saying that he now wishes to illustrate "the *difficulty* of such Attempts." He goes on to describe his attempted reintegrations of amber, vitriol, and turpentine, which he admits to have been only partially successful and apparently not equiponderant.

40. Boyle, *Forms and Qualities*, in Hunter and Davis, *Works*, vol. 5, pp. 355, 371–72.

When he passes to a detailed description of his experiments in the second section of the "Historical Part" of *Forms and Qualities*, he does not, therefore, return to redintegration but gives pride of place to the classical reduction to the pristine state. The first of the ten numbered experiments describes the dissolution of camphor in oil of vitriol to produce a deep yellow-red and almost opaque color despite the fact that both materials were originally colorless. Although the camphor loses its original color and smell in this dissolution, it can be regained in its pristine state (with its piercing odor intact) by merely adding a sufficient quantity of water to the oil of vitriol. Boyle then makes some general comments about this reduction that again reveal the degree of his debt to Sennert and the remarkable inversion that Boyle wrought on his source. After considering the puzzle presented by a specifically lighter body's (the camphor's) submergence in a denser one (the oil of vitriol), Boyle comments on the remarkable color changes that have ensued without the presence of heat or any other agent. He then jumps into a Sennertian disquisition on the significance of the experiment for scholastic theories of mixture:

> This Experiment may serve to countenance what we elsewhere argue against the Schools, touching the Controversie about Mistion. For whereas though some of them dissent, yet most of them maintain, that the Elements alwaies loose their Forms in the mix'd Bodies they constitute; and though if they had dexterously propos'd their Opinion, and limited their Assertions to some cases, perhaps the Doctrine might be tolerated: yet since they are wont to propose it crudely and universally, I cannot but take notice, how little tis favour'd by this Experiment; wherein even a mix'd Body (for such is Camphire) doth, in a further mistion, retain its Form and Nature, and may be immediately so divorced from the Body, to which it was united, as to turn, in a trice, to the manifest Exercise of its former Qualities. And this Experiment being the easiest Instance, I have devis'd, of the preservation of a Body, when it seems to be destroy'd, and of the Recovery of a Body to its former Conditions; I desire it may be taken notice of, as an instance I shall after have Occasion to have recourse to, and to make use of.[41]

First, Boyle points out in Sennertian fashion that the reduction of the camphor reveals the inadequacy of the scholastics' views on mixture, since "most of them" claim that mixture necessitates the loss of the ingredients' forms. Since the camphor is regained intact upon a mere affusion of water, it must have been lurking there all along in the form of

41. Boyle, *Forms and Qualities*, in Hunter and Davis, *Works*, vol. 5, p. 396.

imperceptible corpuscles. These remarks could have come out of Sennert's mouth, but Boyle's following words reveal with great clarity the ease with which the German academic's position could be modified to become that of a mechanical philosopher.

> But the notablest thing in the Experiment is, that Odours should depend so much upon Texture; that one of the subtlest and strongest sented Drugs, that the East it self or indeed the World affords us, should so soon quite loose its Odour, by being mix'd with a Body that has scarce, if at all, any sensible Odour of its own, and This, while the Camphorate Corpuscles survive undestroy'd, in a Liquor, from whence one would think, that lesse subtle and fugitive Bodies, then they, should easily exhale.[42]

Boyle begins his comments here with an observation that might at first seem quite congenial to Sennert's position, for his words can be read in two ways. At first glance, Boyle could be taken as saying simply that the camphor's loss of odor while dissolved in the oil of vitriol must be due merely to a slight change in its texture (rather than resulting from the loss of an immaterial form), since the camphor was not destroyed but regained intact. This would provide an exact parallel to Sennert's discussion of the dissolution and precipitation of silver and other metals, where the major point was that the striking alteration in color and appearance entailed no loss or reacquisition of a form but was due merely to a *diakrisis* and *synkrisis* of atoms. But Boyle does not mean to limit his emphasis on texture to the change from scentless to odoriferous particles. He does not wish to say merely that change in texture produces new qualities but that texture itself is the seat of the quality in question, namely, the odor of the camphor. Hence Boyle ends this section of *Forms and Qualities* with additional modifications to the experiment that allow him to produce a red color from the camphor and to destroy it at will. He then passes from the particular example of the redness, which "appear'd to reside in the Mixture as such" rather than in the individual ingredients, to the general conclusion "that divers of the praeceding *Phaenomena* depend upon the particular Texture of the Liquors, imploy'd to exhibit them." Here once again we can see how the British naturalist has reformulated the Sennertian *reductio in pristinum statum* to become an important weapon in his antihylomorphic arsenal. Once more Sennert's "minima of their own genus" are used to demonstrate the superficial character of the corpuscular texture—brought about by

42. Boyle, *Forms and Qualities*, in Hunter and Davis, *Works*, vol. 5, pp. 396–397.

a mere recess and access of particles—upon which phenomenal qualities in general depend. The impenetrable casings of Sennert's atoms, protective armor sealing in their precious occupants, the substantial forms, had proven to be hollow shells.[43] In short, Sennert's *synkrisis* and *diakrisis*, the operations that were supposed to reveal the permanence of the substantial form, had now become the very tools of its destruction.

43. I do not mean to suggest that Boyle's corpuscular theory was utterly devoid of immaterial causative agents, of course. Antonio Clericuzio has written extensively on Boyle's sporadic references to seminal principles, mainly inherited from earlier chymists such as Severinus and to some degree filtered through the work of Sennert. In my view, however, Boyle's seminal principles were largely a way of reintroducing intelligent agency through the back door in order to explain such difficulties as the propagation of species and extreme cases of regularity in nature—the formation of crystals, for example. In other words, I do not see Boyle's seminal principles as a highly developed part of his system but rather as a sort of rearguard action intended to evade certain explanatory difficulties resulting inevitably from the postulation of a purely mechanical universe. Anstey has come to a position that is not incompatible with mine. See Clericuzio, "A Redefinition of Boyle's Chemistry and Corpuscular Philosophy," *Annals of Science* 47 (1990): 561–589; and Anstey, "Boyle on Seminal Principles," *Studies in History and Philosophy of Biological and Biomedical Sciences* 33 (2002): 597–630. As Anstey says on p. 627, Boyle's seminal principles "were required to explain those phenomena that appeared beyond the capabilities of the corpuscular hypothesis."

A Concise Conclusion

Virtually every survey of the Scientific Revolution highlights the importance of Boyle's mechanical philosophy, but the precise relationship of this doctrine to the immediate matter theory that it replaced has until now received uniformly short shrift. Few historians have appreciated the fact that the mechanical philosophy, as formulated by Robert Boyle, was itself the capstone to a preexisting tradition employing alchemy to recast scholastic theories of mixture, an attempt at reform whose roots extended well into the Middle Ages. Nor does one find a common awareness of the fact that Boyle's most significant experimental evidence for the persistence of microlevel corpuscles and for the mechanical character of the accidental qualities induced upon and removed from those corpuscles stemmed from the reduction to the pristine state originating in the alchemical tradition and made famous in the early seventeenth century by Daniel Sennert.

The radical character of these claims justifies some comment from a methodological perspective. Despite the novelty of its results, the picture that this book paints of medieval and early modern matter theory employs traditional tools of textual scholarship and intellectual history to subvert the complacent story that has become canonical in the existing surveys of the Scientific Revolution. However great the attractions of sociology, anthropology, and critical theory may be to the historian, they cannot supplant the knowledge to be gained from a close reading of the original primary texts with all of their recalcitrant vexations. To eschew a patient and detailed study of the original documents on their own terms invites stagnation and the view that the "real" history, which has supposedly been written, needs only the addition of imported "theoretical" lucubrations to make it complete. But the importation of these postmodernist approaches merely reinforces the modernist assumptions and stereotypes of a previous generation.

My approach will no doubt raise the specter of internalism among some scholars, but to them I reply that it is precisely internalism to which I object. It was, after all, an internalistic vision that led historians such as Butterfield and the Halls to write alchemy out of the picture leading up to the Scientific Revolution. The factors that have made it possible for this book to cast new light on matter theory and the mechanical philosophy are its attention to the precise details of alchemical experimentation, to largely forgotten historical figures who have been excised from the "grand narrative," and to the history of philosophy before and during the seventeenth century, all implicitly excluded by the narrow focus of authors who have accepted the frequently self-serving rhetoric of the heroes of modern science, themselves intent on portraying their discoveries as radically new. In effect, historians have artificially constricted the scope of science in a period when natural philosophy really meant a branch of philosophy as a whole rather than Newtonian physics. This constriction in field has accompanied a parallel telescoping of chronology that has made historians see novelty where gradual development was actually the rule and to see stability where real change was taking place. It is this very internalism that underlies the distorted picture of the Scientific Revolution perpetuated in the contemporary surveys that we examined in the introduction to the present book. Historians can do better, and it is my hope that this book will encourage the efforts of those who are intent upon enriching the story of modern science in its emergence.

The buried edifice of matter theory that we have unearthed provides a window into the remarkable degree of interaction between theory and practice that was possible in medieval and early modern alchemy. It can only be cause for surprise that Paul of Taranto was already making sophisticated use of the reduction to the pristine state to refute Thomistic theories of mixture and the unity of forms and to support a corpuscular theory in the High Middle Ages centuries before Sennert and Boyle made the same move. Equally startling, perhaps, is the way in which early modern thinkers such as Libavius and Sennert yoked Aristotle's meteorology to the matter theory of the alchemists in order to produce a picture of the

Stagirite as a Democritean and the Abderite as an Aristotelian. This move would condition Boyle's early defense of atomism and contribute to his mature corpuscular philosophy even after he had abandoned the atoms of the Greeks. In the meantime, Sennert's Democritizing Aristotle and Aristotelian Democritus had allowed the German philosopher to substitute his own theistic hylomorphism for the starkly materialist ontology of the genuine Democritus in order to protect the place of substantial forms, "the hands of God," in the world of nature. But the gentle transition with which Sennert hoped to elide the passage from scholastic hylomorphism to atomism fell victim to the audacity and shrewdness of Boyle, who seized the discoveries of his fellow corpuscularian Sennert, extracted their marrow, and inserted it into the skeleton of a new theory designed to destroy the very ideas that Sennert held most dear. The features that Sennert added to scholastic hylomorphism—above all, the *emboîtement* of the unchanging essences within "minima of their own genus" and the resultant shift in explanatory power away from formal corruption and replacement to *diakrisis* and *synkrisis*—were just the tools to provide Boyle with his chief means of giving the antihylomorphic mechanical philosophy an experimental basis. For it was precisely such essence-bearing Sennertian corpuscles—not the hypothetical entities of classical atomism incapable of division ab initio, but experimentally attainable bits of matter whose conglomerate texture altered only as a result of relative placement to one another—that Boyle needed in order to illustrate the mechanical character of most phenomenal change and the superfluity of explanation in terms of form. As we have seen, it was Sennert's reductions to the pristine state that allowed Boyle to demonstrate not only the existence of such unchanging corpuscles but also the fact that their accidents could be removed or induced by purely mechanical means.

Despite his unacknowledged debt to Sennert, Boyle's originality and persistence in developing ever more cogent demonstrations of the mechanical philosophy over the course of his career are not to be denied. Boyle's inversion of Sennert's reasoning and his use of the same experimental evidence to yield radically different results bring to mind the parallel case of Antoine Laurent Lavoisier and Joseph Priestly over a century later. As everyone with a nodding acquaintance of the history of science knows, Priestly discovered that red precipitate of mercury (mercury oxide) could be made to give up "dephlogisticated air" (oxygen), but it was Lavoisier who realized the real import of the gas being released and used the phenomenon to reform chemistry from the ground up.[1] How then is Sennert's case any different from Priestly's, where one scientist tried to fit his novel discovery into the framework of an outworn theory while another recognized the latter's threadbare character and built a revolutionary edifice to support his new matter of fact? One could address this question in many ways, but perhaps the most obvious is to reply that Sennert's discovery did not lie in the isolation of a particular material, nor was it really even a matter of fact. What Sennert discovered, rather, was a fatal flaw in the reasoning behind the scholastic hylomorphism of his day—it could not cope in a meaningful way with the reversible reactions made commonplace by early modern chymistry. Instead of simply downplaying or ignoring the problematic evidence coming from chymistry, however, as most of his scholastic contemporaries did, Sennert used it in support of a new atomism that incorporated Aristotelian forms with atoms. Unlike the majority of his peers, Sennert did not reject atoms in favor of Aristotelian matter theory, but instead turned Aristotle into an atomist. He did not see as far as Boyle, but neither did his self-styled atomism look in the opposite direction.

Indeed, it is important to note that Boyle's dream of carrying his microstuctural explanations down to the lowest threshold of material divisibility, his *prima naturalia* or primordial corpuscles, was ultimately unrealizable. It was certainly not this aspect of his work that would influence Lavoisier in the following century but rather Boyle's experimental approach. Lavoisier famously disproved Boyle's attempts to transmute water into earth and improved upon Boyle's experiments with combustion.[2] But the French savant was openly chary of following Boyle into the realm of the microworld. He was content with having employed the analytical tools of the laboratory to

1. The story of Lavoisier's discovery of oxygen is a stock feature of virtually all surveys of the history of chemistry. For a recent example, see Robert Siegfried, *From Elements to Atoms: A History of Chemical Composition* (Philadelphia: American Philosophical Society, 2002), pp. 163–182. For additional background to Lavoisier's famous discoveries, see Henry Guerlac, *Lavoisier— The Crucial Year; The Background and Origin of His First Experiments on Combustion in 1772* (New York: Gordon and Breach, 1961); Frederic Lawrence Holmes, *Antoine Lavoisier—The Next Crucial Year: Or the Sources of His Quantitative Method in Chemistry* (Princeton: Princeton University Press, 1998); and Louise Palmer, "The Early Scientific Work of Antoine Laurent Lavoisier: In the Field and in the Laboratory, 1763-1767" (Ph.D. diss., Yale University, 1998).

2. Like Lavoisier's disproof of phlogiston, these recreations of Boyle's experiments figure in most surveys of the history of chemistry and of science in general. See, for example, Siegfried, *From Elements to Atoms*, pp. 116, 164–165; and A. E. E. McKenzie, *The Major Achievements of Science* (New York: Simon and Schuster, 1973), pp. 91–106.

arrive at fixed chemical species that could be granted the status of elements. It is not hard to see the empirically conservative approach of thinkers like Geber and Sennert in Lavoisier's methods, and as Lawrence Principe and I have argued elsewhere, Lavoisier's emphasis on conservation of weight and on gravimetric techniques as a means of confirming his analyses owe a considerable debt to the chymistry of Joan Baptista Van Helmont, which also had strong roots in corpuscular alchemy.[3]

But one may also push the links between corpuscular alchemy and later chemistry a bit further if we look to the tradition of chemical atomism in the nineteenth century. In 1803, the physical meteorologist John Dalton succeeded in devising a table that linked specific atomic weights to the elementary substances that Lavoisier and his followers had arrived at by means of analysis; this would eventually form the basis of Dalton's famous *New System of Chemical Philosophy* published in 1808. Although Dalton chose to think of the discrete character of atomic weights as reflecting the reality of hard, indivisible atoms, historians of chemistry have shown that many nineteenth-century chemists, while accepting Dalton's contributions to stoichiometry, did not follow him down this path.[4] Alan Rocke, in particular, has made it clear that there were two traditions of atomism current among nineteenth-century chemists. The first, a tradition of "physical atomism" built on the Newtonian tradition that there were "hard, massy unsplittable, impenetrable, spherical atoms," while the second, a parallel school of "chemical atomism," accepted the reality of fixed chemical species having invariant elemental weights but expressed nescience about the ultimate indivisibility of the corpuscles responsible for that invariance.[5] Now it is obvious that pre-Daltonian chymists could

3. Newman and Principe, *Alchemy Tried in the Fire*, pp. 296–309. Unlike many figures in the tradition of alchemical corpuscularism, Van Helmont tends to downplay the *fortissima compositio* of Geber and to contrast the "mere apposition" of corpsucles to the true "wedlock" of substances involved in transmutation per se. Despite this downgrading of Geber's "very strong composition," however, corpuscular ruminations densely populate Van Helmont's writings, and he uses them to explain such varied phenomena as the sublimation of ice, the plating of iron by copper, and the reduction of mercury from its compounds. For Van Helmont's corpuscular theory and its debt to Geberian alchemy, see Newman, *Gehennical Fire*, pp. 141–151.
4. See David M. Knight, *Atoms and Elements: A Study of Theories of Matter in England in the Nineteenth Century* (London: Hutchinson, 1967); William H. Brock, ed., *The Atomic Debates: Brodie and the Rejection of the Atomic Theory; Three Studies* (Leicester: Leicester University Press, 1967); and Alan J. Rocke, *Chemical Atomism in the Nineteenth Century: From Dalton to Cannizarro* (Columbus: Ohio State University Press, 1984).
5. Rocke, *Chemical Atomism*, p. 10.

have had no understanding of atomic weights, since they lacked the new and highly precise analytical data provided by the research of Lavoisier and his followers that allowed Dalton to develop his theory. Yet if we turn our attention from specific atomic weights to the idea of fixed chemical units that represent the limits of analysis but may or may not reflect the ultimate constitution of matter, we are back in a world that would have been most familiar to the tradition of alchemical corpuscular theory in its pre-Boylean phase. The famous explosives chemist Marcelin Berthelot would express agreement in 1877 with "the assumption that all substances consist of very small particles indecomposable by present physical or chemical means."[6] These words, if we dissociate them from Berthelot's following remarks on constant combining proportions, could have come from the mouth of Sennert. As Rocke points out, nineteenth-century chemistry was still heavily indebted to an empiricist tradition stemming from the work of the early eighteenth-century chymist Georg Ernst Stahl that relied on "an operational criterion that elements were to be viewed as simply the last point of analysis." The diffusion of this principle was of paramount importance to the development of subsequent chemistry. As Rocke continues, "the operational criterion of elementarity gradually insinuated itself into the consciousness of chemists, so that by the time Lavoisier first clearly and unambiguously stated it in his classic *Traité élémentaire de chimie*, it could provoke but little controversy."[7] Stahl himself, while in some respects opposed to mechanism, was indebted to the tradition of corpuscular alchemy that resurfaced in the seventeenth century in the works of J. J. Becher and of Eirenaeus Philalethes, the nom de guerre of the American George Starkey.[8] Boyle's mechanical philosophy, in short, was only one of several paths by which the alchemical corpuscularism studied in this book left its traces on posterity. Undoubtedly the further lines of this tradition will receive illumination in subsequent scholarship.[9]

6. Marcelin Berthelot as quoted by Rocke, *Chemical Atomism*, p. 324.

7. Rocke, *Chemical Atomism*, pp. 4–5.

8. Newman, *Gehennical Fire*, pp. 209, 226, 239–242. Unfortunately, a comprehensive study of Stahl remains to be written. His well-known debt to the chymist Johann Joachim Becher is described in Hélène Metzger, *Newton, Stahl, Boerhaave et la doctrine chimique* (Paris: F. Alcan, 1930). See also Irene Strube, *Georg Ernst Stahl* (Leipzig: B.G. Teubner, 1984); and Kevin Chang, "Fermentation, Phlogiston, and Matter Theory: Chemistry and Natural Philosophy in Georg Ernst Stahl's *Zymotechnia Fundamentalis*," *Early Science and Medicine* 7 (2002): 31–64.

9. Lawrence M. Principe is currently composing a study of Wilhelm Homberg and his circle that will very likely cast new light on the fate of chymical corpuscular theory in the eighteenth century.

But what does the story that we have unearthed have to tell us about the way in which the Scientific Revolution should be written? First, it is now beyond doubt that Boyle's experimental program for legitimizing the mechanical philosophy did not stem principally from physics, in the sense that Marie Boas Hall used that term, but from the chymistry of his contemporaries and forebears. This is not to deny the obvious importance of Bacon, Gassendi, and Descartes, whose deep influence on Boyle's thought may be felt in a number of very important areas. Particularly significant for the themes covered in the present study are the facts that Bacon, above all, provided Boyle with the general method of composing experimental histories, inculcated a rigorous experimental methodology, and even supplied him with a highly significant kinetic theory of heat, while Gassendi endowed Boyle with blueprints for specific experiments and Descartes inspired him to his own tentative microstructural explanations in terms of the size, shape, and motion of invisible corpuscles.[10]

But if we look at the empirically based corpuscular reasoning habitually employed by Boyle and his polymorphous usage of the reduction to the pristine state, the paramount role of preexistent chymical theory and practice in establishing the mechanical philosophy cannot be denied. Having acknowledged this, we may then move on to a second point. The tradition of scholastic alchemy, long marginalized as an obscurantist byway in the development of science, now emerges as an essential factor in the development and transmission of experimental corpuscular theory from the High Middle Ages up until the time of Boyle. The alchemy contained in medieval works such as the *Summa perfectionis* of Geber provided the "negative empiricial" approach justifying the Sennertian and Boylean claim that the limits of laboratory analysis also revealed the corpuscular structure of matter. In the course of the present book we have seen the great challenges faced by empirically oriented chymists who wished to assert this principle in the face of scholastic partisans who championed the perfect mixture of Aristotle's *De generatione et corruptione*. But it was precisely the relatively stable corpuscles initially revealed by processes such as sublimation, distillation, and calcination, and later by dissolution in the mineral acids followed by precipitation and reduction, that would provide the backbone of Boyle's mechanical philosophy, not the sterile and unattainable *prima naturalia*

---

10. Sargent, *Diffident Naturalist*, pp. 307, 308, and 310 for the Baconian theory of heat, and passim for Bacon's influence on Boyle overall.

that have traditionally been seen as the sine qua non of this doctrine. Only by looking beyond Boyle's gesturing to the "new philosophy" of his time and by examining the core of his experimental activity in its interaction with theory can we escape the longstanding confusion between what is merely programmatic in his work and what is genuine innovation.

Does the continuity that this book has unearthed, with its surprising revelation that Boyle's mechanical philosophy had an unlikely source in alchemy—even in alchemy's medieval incarnation—imply that the Scientific Revolution is itself a misnomer or a mirage? Is this another self-styled book about a revolution that did not occur?[11] An answer to this question quickly emerges if we consider that the theories and practices employed by Geber and Sennert took their aim against a long-lived and massively popular reading of Aristotle based on the denial of retrievable corpuscles that endured during mixture. The view that genuine mixture could only result when the ingredients of the mixt were destroyed was already established in Thomas Aquinas's thirteenth-century theory that every substance could have only one substantial form, but it reached its widest audience in the printed handbooks and commentaries of the generation directly before Boyle, particularly (though by no means exclusively) in those composed and disseminated by the Jesuits. The theory of perfect mixture and the concomitant denial of its reversibility were iconic features of a conventional scholasticism whose overthrow was genuinely epoch-making. But the equally scholastic corpuscular alchemy inherited and revised by Boyle had always implicitly contained the seeds of hylomorphism's undoing, since its structural explanations suffered an uneasy cohabitation with the peripatetic substantial form. Nonetheless, when Boyle chose to bring this awkwardness to the fore and elected to highlight the advantages of a purely structural type of explanation, he was joining a new battle even if his weapons were not all of his own making. It was Boyle's corpuscular theory that contributed, through Locke, to the movement usually labeled "British empiricism," and it was again Boyle's matter theory that provided the immediate background to Newton's revolutionary discoveries in optics.[12]

11. I have in mind the opening words of Steven Shapin's *The Scientific Revolution* (Chicago: University of Chicago Press, 1996), p. 1: "There was no such thing as the Scientific Revolution, and this is a book about it."

12. For Boyle's influence on Locke, see Peter Alexander, *Ideas, Qualities and Corpuscles* (Cambridge: Cambridge University Press, 1985). The current revival of Boyle studies is sure to

Boyle's ceaseless war on hylomorphism and his reduction of the sensible world to mechanical causes have just as much right to the name of revolution as any political event that results in the deposition of an old and established dynasty, satisfied in its ways and arrogant in its desire to limit discussion to its kingly imperatives while the world moves on. Even if the new regime established by Boyle and his peers soon gave way to Newtonian dynamics, the scholastic "dictatorship of substantial forms" had come to an abrupt and decisive end at the hands of the mechanical philosophers.[13] Like a conquering horde bent on executing its prisoners rather than interrogating them, the mechanical philosophers allowed only a distorted picture of the previous order to remain. One casualty of this philosophical "ethnic cleansing" was Daniel Sennert, whose intermediate role between hylomorphism and mechanism made him an irresistible target both of borrowing and of repudiation. By casting a stronger beam on Sennert's natural philosophy we therefore acquire a better sense of both old and new and begin to see a variety of debts and disjunctures previously undisclosed. Sennert's world was not Boyle's world, despite the fact that the amiable German had provided his philosophical successor with key components of the latter's experimental program to demonstrate the validity of the mechanical philosophy. The irony of Sennert's discomfiture, achieved with weapons of his own devising placed in the hands of an ungrateful heir, is matched only by the schadenfreude that the specter of Sennert might feel if he could encounter the world of modern subatomic physics. Despite the revolutionary character of seventeenth-century mechanism, we should not forget that the microworld, in the end, is just as inhospitable to mechanical philosophers as it is to Aristotelians.[14]

---

lead to a reappraisal of the details linking Boyle and Locke, but the basic fact of Boyle's influence cannot be denied. For Newton's use of Boyle, see Alan Shapiro, *Fits, Passions and Paroxysms: Physics, Method, and Chemistry and Newton's Theories of Colored Bodies and Fits of Easy Reflection* (Cambridge: Cambridge University Press, 1993).

13. The expression "dictatorship of substantial forms" ("Alleinherrschaft der substanziellen Formen") comes from Kurd Lasswitz, *Geschichte der Atomistik* (Hamburg: Leopold Voss, 1890), vol. 1, p. 219. Like Lasswitz, I use these political and military expressions in a metaphorical, rather than a literal sense, of course.

14. I refer to the "familiarity condition" imposed by Boyle's mechanical philosophy, according to which an explanation acquires its force by virtue of explaining the phenomena in terms of more familiar phenomena. As Michael Friedman points out, this criterion does not apply to quarks or many other entities of the modern physicist's world. See Michael Friedman, "Explanation and Scientific Understanding," *Journal of Philosophy* 71 (1974): 5–19.

Bibliography

Aegidius Romanus, Pseudo-. *Commentationes physicae et metaphysicae.* Ursel: Jonas Rhosius, 1604.

Agrippa, Heinrich Cornelius. *De occulta philosophia libri tres.* Ed. V. Perrone Compagni. Leiden: Brill, 1992.

Albertus Magnus. *Beati Alberti Magni . . . opera.* Ed. Pierre Iammy. 21 vols. Lyon: Claudius Prost et al., 1651.

Alexander, Peter. *Ideas, Qualities and Corpuscles.* Cambridge: Cambridge University Press, 1985.

Anstey, Peter. "Boyle on Seminal Principles." *Studies in History and Philosophy of Biological and Biomedical Sciences* 33 (2002): 597–630.

————. *The Philosophy of Robert Boyle.* London: Routledge, 2000.

————. "Robert Boyle and the Heuristic Value of Mechanism." *Studies in History and Philosophy of Science* 33 (2002): 161–174.

Aristotle. *Aristotelis opera cum Averrois commentariis.* 9 vols. Venice: apud Junctas, 1562–1574. Reprint, Frankfurt am Main: Minerva, 1962.

————. *Aristotelis opera edidit academica regia borussica.* 5 vols. Berlin: George Reimer, 1831–1870.

————. *De caelo.* Ed. and trans. W. K. C. Guthrie. Cambridge, MA: Harvard University Press, 1939.

————. *De generatione et corruptione.* Ed. and trans. E. S. Forster. Cambridge, MA: Harvard University Press, 1955.

————. *De generatione et corruptione translatio vetus.* Ed. Joanna Judycka. In *Aristoteles latinus.* Vol. 9.1. Leiden: Brill, 1986.

————. *Metaphysics.* Ed. and trans. Hugh Tredennick. Cambridge, MA: Harvard University Press, 1933.

————. *Météorologiques.* Ed. and trans. Pierre Louis. Paris: Belles Lettres, 1982.

————. *Meteorology.* Ed. and trans. H. D. P. Lee. Cambridge, MA: Harvard University Press, 1952.

————. *Topics.* Ed. and trans. E. S. Forster. Cambridge, MA: Harvard University Press, 1960.

Averroes. In *Aristotelis opera cum Averrois commentariis.* 9 vols. Venice: apud Junctas, 1562–1574. Reprint, Frankfurt am Main: Minerva, 1962.

Avicenna. *Avicennae arabum canon medicinae.* 2 vols. Venice: Iunctae, 1608.

Bacon, Francis. *The Works of Francis Bacon.* Ed. James Spedding, Robert Leslie Ellis, and Douglas Denon Heath. 14 vols. London: Longman, 1857–1874.

Baffioni, Carmela. *Il IV Libro dei "Meteorologica" di Aristotele.* Naples: C.N.R., Centro di studio del pensiero antico, 1981.

Beguin, Jean. *Tyrocinium chymicum recognitum et auctum.* Paris: Matheus le Maistre, 1612.

Bensaude-Vincent, Bernadette, and Isabelle Stengers. *A History of Chemistry.* Cambridge, MA: Harvard University Press, 1996.

Biagioli, Mario. "The Scientific Revolution Is Undead." *Configurations* 6 (1998): 141–148.

Bloch, Olivier René. *La philosophie de Gassendi: Nominalisme, matérialisme et métaphysique.* La Haye: Martinus Nijhoff, 1971.

Boerhaave, Hermann. "De mercurio experimenta." *Philosophical Transactions of the Royal Society of London* [1683–1775], 38 (1733—1734): 145–167.

Bougard, Michel. *La chimie de Nicolas Lemery.* Brepols: Turnhout, 1999.

Boyle, Robert. *The Works of Robert Boyle.* Ed. Michael Hunter and Edward B. Davis. 14 vols. London: Pickering and Chatto, 2000.

Brock, William H., ed. *The Atomic Debates: Brodie and the Rejection of the Atomic Theory, Three Studies.* Leicester: Leicester University Press, 1967.

Butterfield, Herbert. *The Origins of Modern Science, 1300–1800.* New York: Macmillan, 1951.

Butters, Suzanne B. *The Triumph of Vulcan: Sculptors' Tools, Porphyry, and the Prince in Ducal Florence.* Florence: Olschki, 1996.

Cato. *On Agriculture.* Cambridge, MA: Harvard University Press, 1935.

Chalmers, Alan. "Experimental versus Mechanical Philosophy in the Work of Robert Boyle: A Reply to Anstey and Pyle." *Studies in History and Philosophy of Science* 33 (2002): 191–197.

———. "The Lack of Excellence of Boyle's Mechanical Philosophy." *Studies in History and Philosophy of Science* 24 (1993): 541–563.

Chang, Kevin. "Fermentation, Phlogiston, and Matter Theory: Chemistry and Natural Philosophy in Georg Ernst Stahl's *Zymotechnia Fundamentalis*." *Early Science and Medicine* 7 (2002): 31–64.

Clericuzio, Antonio. *Elements, Principles and Corpuscles: A Study of Atomism and Chemistry in the Seventeenth Century*. Dordrecht: Kluwer, 2000.

———. "A Redefinition of Boyle's Chemistry and Corpuscular Philosophy." *Annals of Science* 47 (1990): 561–589.

Cohen, Floris. *The Scientific Revolution: A Historiographical Inquiry*. Chicago: University of Chicago Press, 1994.

Cole, Michael. "Cellini's Blood." *Art Bulletin* 81 (1999): 215–235.

The Compact Edition of the Oxford English Dictionary. Oxford: Oxford University Press, 1971.

Conimbricenses. *Commentarii Collegii Conimbricensis Societatis Iesu, in duos libros de generatione et corruptione Aristotelis*. Lyon: Horatius Cardon, 1606.

Copenhaver, Brian. "Astrology and Magic." In *The Cambridge History of Renaissance Philosophy*. Ed. Charles B. Schmitt and Quentin Skinner, pp. 264–300. Cambridge: Cambridge University Press, 1988.

———. "The Occultist Tradition and Its Critics." In *The Cambridge History of Seventeenth-Century Philosophy*. Ed. Daniel Garber and Michael Ayers, vol. 1, pp. 454–512. Cambridge: Cambridge University Press, 1998.

———. "Scholastic Philosophy and Renaissance Magic in the *De Vita* of Marsilio Ficino." *Renaissance Quarterly* 37 (1984): 523–554.

Daniel, Dane T. "Paracelsus' 'Astronomia Magna' (1537/38): Bible-Based Science and the Religious Roots of the Scientific Revolution." Ph.D. diss., Indiana University, 2003.

Dear, Peter. "The Mathematical Principles of Natural Philosophy: Toward a Heuristic Narrative for the Scientific Revolution." *Configurations* 6 (1998): 173–193.

———. *Revolutionizing the Sciences: European Knowledge and Its Ambitions, 1500–1700*. Princeton: Princeton University Press, 2001.

Debus, Allen G. *Chemistry, Alchemy and the New Philosophy, 1500–1700*. London: Variorum Reprints, 1987.

———. *The Chemical Philosophy: Paracelsian Science and Medicine in the Sixteenth and Seventeenth Centuries.* 2 vols. New York: Science History Publications, 1977.

———. *The French Paracelsians.* Cambridge: Cambridge University Press, 1991.

Dijksterhuis, E. J. *The Mechanization of the World Picture.* Oxford: Oxford University Press, 1961.

Emerton, Norma E. *The Scientific Reinterpretation of Form.* Ithaca, NY: Cornell University Press, 1984.

Erastus, Thomas. *Explicatio quaestionis famosae illius, utrum ex metallis ignobilioribus aurum verum & naturale arte conflari possit.* Appendix to *Disputationes de nova Philippi Paracelsi medicina.* Basel: Petrus Perna, 1572.

Frank, Robert. *Harvey and the Oxford Physiologists.* Berkeley: University of California Press, 1980.

Friedensburg, Walter. *Geschichte der Universität Wittenberg.* Halle: Max Niemeyer, 1917.

Friedman, Michael. "Explanation and Scientific Understanding." *Journal of Philosophy* 71 (1974): 5–19.

Furley, David. "The Mechanics of Meteorologica IV. A Prolegomenon to Biology." In *Zweifelhaftes im Corpus Aristotelicum.* Ed. Paul Moraux and Jürgen Wiesner, pp. 73–93. Berlin: de Gruyter, 1983.

Galen, *Mixture.* In *Galen: Selected Works.* Trans. P. N. Singer. Oxford: Oxford University Press, 1997.

Gantenbein, Urs Leo. *Der Chemiater Angelus Sala.* Zurich: Juris Druck & Verlag Dietikon, 1992.

Gassendi, Pierre. *Animadversiones in decimum librum Diogenis Laertii.* Lyon: Guillaume Barbier, 1649.

———. *Petri Gassendi Dinensis . . . opera omnia.* Lyon: Laurentius Anisson & Ioannes Baptista Devenet, 1658.

Goclenius, Rudolph. *Conciliator philosophicus.* Kassell: Officina Mauritiana, 1609. Olms reprint, 1977.

———. *Lexicon philosophicum.* Frankfurt: Petrus Musculus, 1613. Olms reprint, 1964.

Golinski, Jan. *Making Natural Knowledge.* Cambridge: Cambridge University Press, 1998.

Gregory, Tullio. "Studi sull'atomismo del seicento II." *Giornale critico della filosofia italiana* 45 (1966): 44–63.

Guerlac, Henry. *Lavoisier—The Crucial Year: The Background and Origin of His First Experiments on Combustion in 1772.* New York: Gordon and Breach, 1961.

Guibert, Nicholas. *Alchymia ratione et experientia ita demum viriliter impugnata.* Strasbourg: Zetzner, 1603.

Haas, Frans A. J. de. "Mixture in Philoponus. An Encounter with a Third Kind of Potentiality." In *The Commentary Tradition on Aristotle's "De generatione et corruptione."* Ed. J. M. M. H. Thijsen and H. A. G. Brakhuis, pp. 21–46. Turnhout: Brepols, 1999.

Hall, A. Rupert. *The Scientific Revolution, 1500–1800: The Formation of a Modern Scientific Attitude.* Boston: Beacon Press, 1962.

[Hall], Marie Boas. "The Establishment of the Mechanical Philosophy." *Osiris* 10 (1952): 412–541.

———. *The Scientific Renaissance, 1450–1630.* New York: Harper Torchbooks, 1962.

———. *Robert Boyle and Seventeenth-Century Chemistry.* Cambridge: Cambridge University Press, 1958.

Halleux, Robert. *Les textes alchimiques.* Brepols: Turnhout, 1979.

Halleux, Robert, and Anne-Catherine Bernès. "La cour savant d'Ernest de Bavière." *Archives internationales d'histoire des sciences* 45 (1995): 3–29.

Hannaway, Owen. *The Chemists and the Word: The Didactic Origins of Chemistry.* Baltimore: Johns Hopkins University Press, 1975.

———. "Laboratory Design and the Aim of Science: Andreas Libavius versus Tycho Brahe." *Isis* 77 (1986): 585–610.

Henry, John. *The Scientific Revolution and the Origins of Modern Science.* New York: St. Martin's Press, 1997.

———. Review of Peter Dear, *Revolutionizing the Sciences.* In *British Journal for the History of Science* (June 2004): 199–200.

Héron d'Alexandrie. *Les mécaniques ou l'élévateur des corps lourds.* Ed. and trans. B. Carra de Vaux. Paris: Belles Lettres, 1988.

Hirai, Hiroshi. "Le concept de semence dans les théories de la matière à la Renaissance." 2 vols. Ph.D. diss, Université Lille III, 1999.

Holmes, Frederic Lawrence. *Antoine Lavoisier—The Next Crucial Year: Or the Sources of His Quantitative Method in Chemistry.* Princeton: Princeton University Press, 1998.

Hooykaas, Reijer. "The Experimental Origin of Chemical Atomic and Molecular Theory Before Boyle." *Chymia* 2 (1949): 65–80.

Hunter, Michael. "How Boyle Became a Scientist." *History of Science* 33 (1995): 59–103.

Jacquart, Danielle. "Minima in Twelfth-Century Medical Texts from Salerno." In *Medieval and Early Modern Corpuscularian Matter Theories.* Ed. Christoph

Lüthy, John E. Murdoch, and William R. Newman, pp. 39–56. Leiden: Brill, 2001.

Jessenius, Johann. *De morbi, quem aer tota substantia noxius peragit, praeservatione & curatione disputatio IV. Quam peculiari collegio, praeside Iohan. Iessenio a Iessen, Doctore & Professore. Ad Cal. Septembris adornat Daniel Sennert Vratislaviensis Sil.* Wittenberg: Iohannes Dörffer typis Cratonianis, 1596.

———. *Iohan. Iessenii a Iessen De sympathiae et antipathiae rerum naturalium causis disquisitio singularis. Quam in publico pro virili ad Cal. Iunij defendere conabitur M. Daniel Sennertus Vratislaviensis.* Wittenberg: Meißner, 1599.

Kahn, Didier. "Paracelsisme et alchimie en France à la fin de la Renaissance (1567–1625)." Ph.D. thesis, Université de Paris IV, 1998.

Kangro, Hans. "Erklärungswert und Schwierigkeiten der Atomhypothese und ihrer Anwendung auf chemische Probleme in der ersten Hälfte des 17. Jahrhunderts." *Technikgeschichte* 35 (1968): 14–36.

———. *Joachim Jungius' Experimente und Gedanken zur Begründung der Chemie als Wissenschaft.* Wiesbaden: Franz Steiner, 1968.

Kargon, Robert Hugh. *Atomism in England from Hariot to Newton.* Oxford: Clarendon Press, 1966.

Kaufmann, Thomas DaCosta. "Kunst und Alchemie." In *Moritz der Gelehrte: Ein Renaissancefürst in Europa.* Ed. Heiner Borggrefe et al., pp. 370–377. Eruasberg: Minerva, 1997.

Kirk, G. S., J .E. Raven, and M. Schofield. *The Presocratic Philosophers: A Critical History with a Selection of Texts.* Cambridge: Cambridge Univerity Press, 1983.

Knight, David M. *Atoms and Elements: A Study of Theories of Matter in England in the Nineteenth Century.* London: Hutchinson, 1967.

Kubbinga, Henk. *L'histoire du concept de 'molécule'.* Berlin: Springer, 2002.

Kuhn, Thomas, "Robert Boyle and Structural Chemistry." *Isis* 43 (1952): 12–36.

Lasswitz, Kurd. *Geschichte der Atomistik.* 2 vols. Hamburg: Leopold Voss, 1890.

Lavoisier, Antoine Laurent. *Elements of Chemistry.* Trans. R. Kerr. Edinburgh, 1790.

Lennox, James G. "Robert Boyle's Defense of Teleological Inference in Experimental Science" *Isis* 74 (1983): 38–53.

Libavius, Andreas. *Alchymia . . . recognita, emendata, et aucta.* Frankfurt: Petrus Kopffius, 1606.

———. *Alchymia triumphans.* Frankfurt: Petrus Kopffius, 1607.

———. *Defensio et declaratio perspicua alchymiae transmutatoriae.* Ursel: Petrus Kopffius, 1604.

———. *De mundi corporumque mixtorum elementis et principiis platonicis, aristotelicis, hippocraticis . . . disputandum proposita in illustri Casimiriano Saxonico Coburgi praeside Andrea Libavio M.D. doctore & professore. respondente Alberto Theodorico Thurnaviensi fr. studioso publico nono Iulii anno. 1608. horis matutinis in auditorio inferiore.* Coburg: Iustus Hauck, 1608.

———. *Rerum chymicarum epistolica forma liber primus.* Frankfurt: Petrus Kopffius, 595.

Lohr, Charles. *Latin Aristotle Commentaries.* Florence: Olschki, 1988.

Lucretius. *De rerum natura,* Ed. and trans. W. H. D. Rouse. Cambridge, MA: Harvard University Press, 1937.

Lüthy, Christoph. "An Aristotelian Watchdog as Avant-garde Physicist: Julius Caesar Scaliger." *Monist* 84 (2001): 542–561.

———. "The Fourfold Democritus on the Stage of Early Modern Science." *Isis* 91 (2000): 443–479.

Lüthy, Christoph, and Willam R. Newman. "Daniel Sennert's Earliest Writings (1599–1600) and Their Debt to Giordano Bruno." *Bruniana and Campanelliana* 6 (2000/2): 261–279.

Lüthy, Christoph, Cees Leijenhorst, and Johannes M. M. H. Thijssen. "The Tradition of Aristotelian Natural Philosophy. Two Theses and Seventeen Answers." In *The Dynamics of Aristotelian Natural Philosophy from Antiquity to the Seventeenth Century.* Ed. Cees Leijenhorst, Christoph Lüthy, and Joahnnes M. M. H. Thijssen, pp. 1–29. Leiden: Brill, 2002.

Maier, Anneliese. *An der Grenze von Scholastik und Naturwissenschaft.* 2. Auflage. Rome: Edizioni di Storia e Letteratura, 1952.

———. *Die Mechanisierung des Weltbilds im 17. Jahrhundert.* Leipzig: Felix Meiner, 1938.

Mandelbaum, Maurice. *Philosophy, Science, and Sense Perception.* Baltimore: Johns Hopkins Press, 1964; paperback edition, 1966.

Martin, Craig. *Interpretation and Utility: The Renaissance Commentary Tradition on Aristotle's* "Meteorologica IV." Ph.D. diss., Harvard University, 2002.

Matton, Sylvain. "Les théologiens de la Compagnie de Jésus et l'alchimie." In *Aspects de la tradition alchimique au XVIIᵉ siècle.* Ed. Frank Greiner, pp. 383–501. Paris: S.É.H.A., 1998.

McKenzie, A. E. E. *The Major Achievements of Science.* New York, Simon and Schuster, 1973.

McMullin, Ernan. "Structural Explanation." *American Philosophical Quarterly* 15 (1978): 139–147.

Meinel, Christoph. "Early Seventeenth-Century Atomism: Theory, Epistemology, and the Insufficiency of Experiment." *Isis* 79 (1988): 68–103.

———. "In physicis futurum saeculum respicio." *Veröffentlichung der Joachim Jungius-Gesellschaft der Wissenschaften*. Göttingen: Vandenhoeck & Ruprecht, 1984.

Melsen, Andreas van. *Atom Gestern und Heute*. Freiburg: Alber, 1957.

Metzger, Hélène. *Newton, Stahl, Boerhaave et la doctrine chimique*. Paris: F. Alcan, 1930.

Michael, Emily. "Daniel Sennert on Matter and Form: At the Juncture of the Old and the New." *Early Science and Medicine* 2 (1997): 272–299.

———. "Sennert's Sea Change: Atoms and Causes." In *Late Medieval and Early Modern Corpuscular Matter Theories*. Ed. Christoph Lüthy, John E. Murdoch, and William R. Newman, pp. 331–363. Leiden: E.J. Brill, 2001.

Moran, Bruce. *The Alchemical World of the German Court*. Stuttgart: Franz Steiner, 1991.

———. *Distilling Knowledge: Alchemy, Chemistry, and the Scientific Revolution*. Cambridge, MA: Harvard University Press, 2005.

———. More, Henry. *The Immortality of the Soul*. London: William Morden, 1659.

Multhauf, Robert. "The Relationship between Technology and Natural Philosophy, ca. 1250–1650: As Illustrated by the Technology of the Mineral Acids." Ph.D. diss., University of California, 1953.

Newman, William R. "The Alchemical Sources of Robert Boyle's Corpuscular Philosophy." *Annals of Science* 53 (1996): 567–585.

———. "Alchemical Symbolism and Concealment: The Chemical House of Libavius." In *The Architecture of Science*. Ed. Peter Galison and Emily Thompson, pp. 59–77. Cambridge, MA: MIT Press, 1999.

———. "Alchemy, Assaying, and Experiment." In *Instruments and Experimentation in the History of Chemistry*. Ed. Frederic L. Holmes and Trevor H. Levere, pp. 35–54. Cambridge, MA: MIT Press, 2000.

———. "'Decknamen or pseudochemical Language'? Eirenaeus Philalethes and Carl Jung." *Revue d'histoire des sciences* 49 (1996): 159–188.

———. "Experimental Corpuscular Theory in Aristotelian Alchemy: From Geber to Sennert." In *Late Medieval and Early Modern Corpuscular Matter Theories*. Ed. Christoph Lüthy, John E. Murdoch, and William R. Newman, pp. 291–329. Leiden: E.J. Brill, 2001.

———. *Gehennical Fire: The Lives of George Starkey, an American Alchemist in the Scientific Revolution*. Chicago: University of Chicago Press, 2003 (first published in 1994).

_____. "The Genesis of the *Summa perfectionis*." *Les archives internationales d'histoire des sciences* 35 (1985): 240–302.

_____. "New Light on the Identity of Geber." *Sudhoffs Archiv für die Geschichte der Medizin und der Naturwissenschaften* 69 (1985): 76–90.

_____. *Promethean Ambitions: Alchemy and the Quest to Perfect Nature.* Chicago: University of Chicago Press, 2004.

_____. "Robert Boyle's Debt to Corpuscular Alchemy." In *Robert Boyle Reconsidered.* Ed. Michael Hunter, pp. 107–118. Cambridge: Cambridge University Press, 1994.

_____. "The *Summa perfectionis* and Late Medieval Alchemy." 4 vols. Harvard University, Ph.D. diss., 1986.

_____. *The 'Summa Perfectionis' of Pseudo-Geber: A Critical Edition, Translation and Study.* Leiden: E. J. Brill, 1991.

Newman, William R., and Lawrence M. Principe. *Alchemy Tried in the Fire: Starkey, Boyle, and the Fate of Helmontian Chymistry.* Chicago: University of Chicago Press, 2002.

_____. "Alchemy vs. Chemistry: The Etymological Origins of a Historiographic Mistake." *Early Science and Medicine* 3 (1998): 32–65.

Newton, Isaac. *Opticks.* New York: Dover, 1952.

Nifo, Agostino. *In Aristotelis libros metaphysices.* Venice: Hieronymus Scotus, 1559; Minerva reprint, 1967.

Nummedal, Tara. "Practical Alchemy and Commercial Exchange in the Holy Roman Empire." In *Merchants and Marvels: Commerce, Science, and Art in Early Modern Europe.* Ed. Pamela H. Smith and Paula Findlen, pp. 201–222. New York: Routledge, 2002.

Nutton, Vivian. "The Reception of Fracastoro's Theory of Contagion: The Seed That Fell Among Thorns?" *Osiris*, 2nd ser. 6 (1990): 196–234.

_____. "The Seeds of Disease: An Explanation of Contagion and Infection from the Greeks to the Renaissance." *Medical History* 27 (1983): 1–34.

Osler, Margaret. Review of Peter Dear, *Revolutionizing the Sciences.* In *Annals of Science* 61 (2004): 134–136

Osler, Margaret J., ed. *Rethinking the Scientific Revolution.* Cambridge: Cambridge University Press, 1999.

Pagel, Walter. *Paracelsus: An Introduction to Philosophical Medicine in the Era of the Renaissance.* Basel: Karger, 1982.

Palmer, Louise. "The Early Scientific Work of Antoine Laurent Lavoisier: In the Field and in the Laboratory, 1763–1767." Ph.D. diss., Yale University, 1998.

Paracelsus, Theophrastus, *Theophrastus von Hohenheim, genannt Paracelsus, Sämtliche Werke, I. Abteilung.* Ed. Karl Sudhoff. 14 vols. Munich: Oldenbourg, 1922–1933.

Patterson, T. S. "Jean Beguin and His *Tyrocinium chymicum.*" *Annals of Science* 2 (1937): 243–298.

Pfeifer, Xaver. *Die Controverse über das Beharren der Elemente in den Verbindungen von Aristoteles bis zur Gegenwart, Programm zum Schlusse des Studienjahrs 1878/79.* Dillingen: Adalbert Kold, 1879.

Pick, Friedel. *Joh. Jessenius de Magna Jessen, Arzt und Rektor in Wittenberg und Prag hingerichtet am 21. Juni 1621, in Studien zur Geschichte der Medizin.* Vol. 15. Leipzig: Ambrosius Barth, 1926.

Pliny, *Natural History.* Cambridge, MA: Harvard University Press, 1968.

Principe, Lawrence M. *The Aspiring Adept: Robert Boyle and His Alchemical Quest.* Princeton: Princeton University Press, 1998.

———. "Boyle's Alchemical Writings: Anonymity, Uncertainty, and Oblivion." Paper presented at annual History of Science Society conference, 1993.

———. "Diversity in Alchemy: The Case of Gaston 'Claveus' DuClo, a Scholastic Mercurialist Chrysopoeian." In *Reading the Book of Nature: The Other Side of the Scientific Revolution.* Ed. Allen G. Debus and Michael T. Walton, pp. 181–198. Kirksville: Sixteenth-Century Journal, 1998.

———. "Newly Discovered Boyle Documents in the Royal Society Archive." *Notes and Records of the Royal Society* 49 (1995): 57–70.

———. "Style and Thought of the Early Boyle: Discovery of the 1648 Manuscript of *Seraphic Love.*" *Isis* 85 (1994): 247–260.

———. "Virtuous Romance and Romantic Virtuoso: The Shaping of Robert Boyle's Literary Style." *Journal of the History of Ideas* 56 (1995): 377–397.

Principe, Lawrence M., and Lloyd DeWitt. *Transmutations: Alchemy in Art. Selected Works from the Eddleman and Fisher Collections at the Chemical Heritage Foundation.* Philadelphia: Chemical Heritage Foundation, 2002.

Principe, Lawrence M., and William R. Newman. "Some Problems with the Historiography of Alchemy." In *Secrets of Nature: Astrology and Alchemy in Early Modern Europe.* Ed. William R. Newman and Anthony Grafton, pp. 385–431. Cambridge, MA: MIT Press, 2001.

Pyle, Andrew. *Atomism and Its Critics.* Bristol: Thoemmes Press, 1995.

———. "Boyle on Science and the Mechanical Philosophy: A Reply to Chalmers." *Studies in History and Philosophy of Science* 33 (2002): 175–190.

Ramsauer, Rembert. *Die Atomistik des Daniel Sennert: Als Ansatz zu einer deutschartig-schauenden Naturforschung und Theorie der Materie im 17. Jahrhundert.* Braunschweig: Vieweg, 1935.

Rees, Graham. "Francis Bacon's Semi-Paracelsian Cosmology." *Ambix* 22 (1975): 81–101.

Reif, Sister Mary Richard. "Natural Philosophy in Some Early Seventeenth Century Scholastic Textbooks." Ph.D. diss., St. Louis University, 1962.

Richardson, Linda Deer. "The Generation of Disease: Occult Causes and Diseases of the Total Substance." In *The Medical Renaissance of the Sixteenth Century*, pp. 175–194. Cambridge: Cambridge University Press, 1985.

Rocke, Alan J. *Chemical Atomism in the Nineteenth Century: From Dalton to Cannizarro*. Columbus: Ohio State University Press, 1984.

Rossi, Paolo. *Francesco Bacone: Dalla magia alla scienza*. Bari: Laterza, 1957.

Rudolph, Hartmut. "Hohenheim's Anthropology in the Light of His Writings on the Eucharist." In *Paracelsus: The Man and His Reputation, His Ideas and Their Transformation*. Ed. Ole Peter Grell, pp. 187–206. Leiden: Brill, 1998.

Rütten, Thomas. *Demokrit, lachender Philosoph und sanguinischer Melancholiker: Eine pseudohippokratische Geschichte*. Leiden: Brill, 1992.

Sargent, Rose-Mary. *The Diffident Naturalist: Robert Boyle and the Philosophy of Experiment*. Chicago: University of Chicago Press, 1995.

Scaliger, Julius Caesar. *Exotericarum exercitationum liber xv*. Lyon: Vidua Antonii de Harsy, 1615.

Sennert, Daniel. *Danielis Sennerti Vratislaviensis ... operum in sex tomos divisorum*. 6 vols. Lyon: Joannes Antonius Huguetan, 1676.

———. *De chymicorum cum Aristotelicis et Galenicis consensu ac dissensu*. Wittenberg: Schürer, 1619.

———. *De chymicorum cum Aristotelicis et Galenicis consensu ac dissensu*. Wittenberg: Schürer, 1629.

———. *De differentiis morborum disputatio prima. Cujus theses, cum deo, sub praesidio Danielis Sennerti, phil. et medic. doctoris et profess. p. defendendas suscepit Martinus Boecherus Austriacus. Ad diem 19. Ianuarij*. Wittenberg: Johann Schmidt, 1605.

———. *De febrium malignarum natura & causis disputatio; quam, cum deo sub praesidio Danielis Sennerti d. et med. profess. pub. publico examini subjicit M. Michael Döring, Vratislaviensis, in auditorio medicorum, Ad diem 13. Februarij*. Wittenberg: Martin Henckelius, 1607.

———. *De methodo medendi disputatio vi. de purgatione: in qua, cum deo, sub praesidio Danielis Sennerti, medic. doct. et profess. p. respondentis partes sustinebit Conradus Schattenbergius, Flenspurgensis Holsatus. ad diem 31. Martij*. Wittenberg: Crato, 1604.

———. *De pestilentia disputatio: ad cujus theses, cum deo, praeside Daniele Sennerto, Vratislaviensi, d. & medicinae professore publ. respondebit M. Timotheus Ulricus*

Torgensis pro licentia doctoris in arte medica gradum consequendi: ad diem 6 Novem. horis & loco consuetis. Wittenberg: Martin Henckelius, 1607.

―――. *Epitome naturalis scientiae.* Wittenberg: Caspar Heiden, 1618.

―――. *Epitome naturalis scientiae, comprehensa disputationibus viginti sex.* Wittenberg: Simon Gronenberg, 1600.

―――. *Hypomnemata physica.* Frankfurt: Schleichius, 1636.

―――. *Institutionum medicinae libri v.* Wittenberg: Schürer, 1611.

Severinus, Petrus. *Idea medicinae.* Basil: Henricpetrus, 1571.

Shackelford, Jole. *A Philosophical Path for Paracelsian Medicine: The Ideas, Intellectual Context, and Influence of Petrus Severinus, 1540–1602.* Copenhagen: Museum Tusculanum Press, 2004.

―――. "Seeds with a Mechanical Purpose: Severinus' Semina and Seventeenth-Century Matter Theory." In *Reading the Book of Nature: The Other Side of the Scientific Revolution.* Ed. Allen G. Debus and Michael T. Walton, pp. 15–44. Kirksville: Sixteenth Century Journal Publishers, 1998.

Shanahan, Timothy. "Teleological Reasoning in Boyle's *Disquisition about Final Causes.*" In *Robert Boyle Reconsidered.* Ed. Michael Hunter, pp. 177–192. Cambridge: Cambridge University Press, 1994.

Shapin, Steven. *The Scientific Revolution.* Chicago: University of Chicago Press, 1996.

―――. "Social Uses of Science," In *The Ferment of Knowledge.* Ed. G. S. Rousseau and Roy Porter, pp. 93–139. Cambridge: Cambridge University Press, 1980.

Shapiro, Alan. *Fits, Passions and Paroxysms: Physics, Method, and Chemistry and Newton's Theories of Colored Bodies and Fits of Easy Reflection.* Cambridge: Cambridge University Press, 1993.

Siegfried, Robert. *From Elements to Atoms: A History of Chemical Composition.* Philadelphia: American Philosophical Society, 2002.

Siraisi, Nancy. "Giovanni Argenterio and Sixteenth-Century Medical Innovation: Between Princely Patronage and Academic Controversy" *Osiris,* 2nd ser., 6 (1990): 161–180.

Smith, Pamela H. *The Body of the Artisan: Art and Experience in the Scientific Revolution.* Chicago: University of Chicago Press, 2004.

―――. *The Business of Alchemy: Science and Culture in the Holy Roman Empire.* Princeton: Princeton University Press, 1994.

Staden, Heinrich von. "Teleology and Mechanism: Aristotelian Biology and Early Hellenistic Medicine." In *Aristotelische Biologie.* Ed. Wolfgang Kullmann and Sabine Föllinger, pp. 183–208. Stuttgart: Franz Steiner, 1996.

Stahl, Daniel. *Axiomata philosophica sub titulis xx. Comprehensa.* Cambridge: Roger Daniel, 1645.

Starkey, George. *Alchemical Laboratory Notebooks and Correspondence.* Ed. William R. Newman and Lawrence M. Principe. Chicago: University of Chicago Press, 2004.

Sternagel, Peter. *Die Artes Mechanicae im Mittelalter: Begriffs- und Bedeutungsgeschichte bis zum Ende des 13. Jahrhunderts,* Münchener Historische Studien, Abteilung Mittelalterliche Geschichte, Herausgegeben von Johannes Spürl, Band II. Kallmünz über Regensburg: Michael Lassleben, 1966.

Stolberg, Michael. "Particles of the Soul: The Medical and Lutheran Context of Daniel Sennert's Atomism," in *Medicina nei secoli* 15 (2003): 177–203.

Strohm, H. "Beobachtungen zum vierten Buch der aristotelischen Meteorologie." In *Zweifelhaftes im Corpus Aristotelicum.* Ed. Paul Moraux and Jürgen Wiesner, pp. 94–115. Berlin: de Gruyter, 1983.

Subow, W. "Zur Geschichte des Kampfes zwischen dem Atomismus und dem Aristotelismus im 17. Jahrhundert (Minima naturalia und Mixtio)." In *Sowjetische Beiträge zur Geschichte der Naturwissenschaften.* Ed. Gerhard Harig, pp. 161–191. Berlin, 1960.

Telle, Joachim. *Analecta Paracelsica.* Stuttgart: Franz Steiner, 1994.

———. *Parerga Paracelsica.* Stuttgart: Franz Steiner, 1991.

Telle, Joachim, and Wilhelm Kühlmann. *Corpus Paracelsisticum: Dokumente frühneuzeitlicher Naturphilosophie in Deutschland.* Tübingen: Niemeyer, 2001.

Thackray, Arnold. *Atoms and Powers.* Cambridge, MA: Harvard University Press, 1970.

Thomas Aquinas. *Opera omnia curante Roberto Busa S.I.* Stuttgart, 1980.

———. *Sancti Thomae Aquinatis doctoris angelici opera omnia.* Rome: Typographia Polyglotta, 1882–.

———. *Sancti Thomae Aquinatis doctoris angelici ordinis praedicatorum opera omnia.* 25 vols. Parma: Petrus Fiaccadorus, 1852–1873.

Thorndike, Lynn. *A History of Magic and Experimental Science.* 8 vols. New York: Columbia University Press, 1923–1958.

Toletus, Franciscus. *Francisci Toleti Societatis Iesu, Nunc S.R.E. Cardinalis Ampliss. Commentaria, una cum quaestionibus, in duos libros Aristotelis, de generatione & corruptione, nunc denuo in lucem edita, ac diligentius emendata.* Venice: Iuntae, 1602.

Van Helmont, Joan Baptista. *Opuscula medica inaudita.* Amsterdam: Elsevier, 1648.

Webster, Charles. *The Great Instauration.* New York: Holmes & Meier, 1976.

Weeks, Andrew. *Paracelsus: Speculative Theory and the Crisis of the Early Reformation.* Albany: State University of New York Press, 1997.

Westfall, Richard. *The Construction of Modern Science: Mechanisms and Mechanics.* Cambridge: Cambridge University Press, 1977.

Yates, Frances. "The Hermetic Tradition in Renaissance Science." In *Art, Science, and History in the Renaissance.* Ed. Charles S. Singleton, pp. 255–274. Baltimore: Johns Hopkins University Press, 1967.

———. *The Rosicrucian Enlightenment.* London: Routledge and Kegan Paul, 1972.

Young, John T. *Faith, Medical Alchemy and Natural Philosophy: Johann Moriaen, Reformed Intelligencer and the Hartlib Circle.* Brookfield, VT: Ashgate, 1998.

Zabarella, Jacobus. *Jacobi Zabarellae Patavini, de rebus naturalibus libri xxx.* Frankfurt: Lazarus Zetzner, 1607. Minerva reprint, 1966.

Zavalloni, Roberto. *Richard de Mediavilla et la controverse sur la pluralité des formes.* Louvain: Éditions de l'institut supérieur de philosophie, 1951.

Index

Abderite. *See* Democritus

Access and recess, 182, 184, 189, 203, 215. See also *Synkrisis* and *diakrisis*

Accident, 36, 62, 103, 148, 209, 217; *accidens praedicabile*, 199–200; *accidens praedicamentale*, 199–200

Acids, mineral, 42, 98, 102, 123, 132, 135, 145, 150, 152, 170, 193, 198, 210, 223. *See also* Menstruum; Saline spirits

Act and potency, 49, 75, 78, 102, 149–150

Aegidius Romanus (pseudo), 119–120

Affinity, 134–135, 138

Aggregate corpuscles, 172–173, 179–180, 182, 185, 188, 191–192, 195–215

Agrippa von Nettesheim, Heinrich Cornelius, 77

Albertus Magnus, 16, 26, 28

Alcohol, 72

Alembic, 90, 127, 132

Alexander of Aphrodisias, 118

Alkali, 72

Aludel, 29

Amalgam, 173

Amber, 212

Analysis, 34, 41, 44–46, 49, 57–58, 65, 67–68, 80, 97–99, 144, 166, 191, 208, 210, 220; equiponderant, 211–212; gravimetric, 211, 221; limit of, xi–xii, 30, 45, 60, 127, 144, 161, 222; and synthesis, 67–68, 210–212. *See also* atom; negative-empirical principle

Anaxagoras, 76

Anstey, Peter, 177

Antimony, 144, 173–174

Antipathy, 135, 140, 144; between niter and sulfur, 135; between oil and water, 135–136

Apples, 204

Aqua fortis. See nitric acid

Aqua regia, 98, 150, 173–174, 201, 203

Aquinas, Thomas, and Thomists, 5, 16, 20, 35–38, 50, 53, 58, 64, 102, 107–108, 110–111, 116–118, 124, 150, 152, 167, 224

Arcana maiora, 181

Aristotle: and Aristotelians, 6, 13, 17, 27, 30–31, 52, 55, 58, 60, 66, 69–73, 75, 79, 81, 86–88, 91–94, 107, 117, 139, 153, 168, 173, 189, 193, 195,

209, 219, 225; biology, 68; works of:
De anima, 40, 56; *De caelo*, 25, 49, 76,
97; *De generatione et corruptione*, 14,
25, 27–28, 30–31, 43–44, 48, 50,
52–53, 67–68, 100–101, 107, 110,
115–116, 119, 121, 131, 137,
163–164, 193; 223; *Metaphysics*, 25,
50, 56–57, 60, 75, 199; *Meteorology*,
13–14, 19, 28, 65–68, 70–73, 75,
78–79, 93–94, 96; *Physics*, 25, 114
Arnald of Villanova, 43
Ars mechanica, 185
Art, 63; ape of nature, 62, 64; cannot
reproduce works of nature by its own
power, 88, 114; limited power of its
analysis, 127; versus nature, 48
Assaying tests, 33–34
Association and dissociation, 70, 75–76,
93
Atom, 28, 90–98, 121–124, 127–128,
130–136, 147, 152, 160–170, 200,
214; as last point of analysis, xi–xii,
97, 127; mathematical, xi; multiple
meanings of, xi–xii; of light, 169; as a
perfect mixt, 75–78, 80–81, 92–94,
163; position and arrangement of,
133, 152; shape of, 129, 131, 152;
separated by void, 24; steams of, 170.
See also *metakinēsis; taxis; thesis*
Atomic weight, 221
Atomism, xi–xii, 24, 34, 43–44, 68–70,
74–75, 85–153, 160–170, 218;
chemical atomism, 46, 97, 221–222
Augustine, Sait, 77
Averroes and Averroists, 57, 100, 110,
115, 118, 124, 148, 150, 167
Avicenna, 17, 53, 137; claim that
elements remain intact in a mixture,
95; *qualitates fractae*, 95; on
unknowability of substantial forms,
137

Bacon, Francis, 1, 12, 14, 24, 123, 157,
159, 178, 181, 190, 206, 208, 223; and
heat, 203

Bacon, Roger, 26, 50
Balance, 205
Bauhin, Caspar, 71
Beguin, Jean, 79; *Tyrocinium chemicum*,
79
Bell, 196
Bensaude-Vincent, Bernadette, 96
Bernardus Trevirensis, 43
Berzelius, J. Jacob, 5
Bethelot, Marcelin, 222
Blood, 49, 51, 132, 163, 165, 195
Boerhaave, Hermann, 34
Boyle, Robert, 1–3, 5–9, 11–16, 18–20,
23–24, 44, 77, 96–97, 152–153,
157–215, 217–225; and definition of
an element, 96; reticence, 204; and
seminal principles, 77; strategies for
demonstrating mechanical origin of
qualities, 188–189, 194, 196–198,
200–215; works of: *Certain
Physiological Essays*, 210; "Essay of
Salt-Petre," 210, 212; *Essay of the holy
Scriptures*, 1; *Experiments and
Considerations Touching Colours*,
182–185, 198; *Experiments, Notes, & c.
about the Mechanical Origine or
Production of Divers Particular
Qualities*, 162, 201; "Free
Consideratons about Subordinate
Formes," 210; "History of Fluidity,"
204; "History of Particular
Qualities," 162, 195–197, 203,
206–207; *Of the Atomicall Philosophy*,
160–170, 172–173, 175, 191; *Origine
of Formes and Qualities*, 1, 20, 167, 182,
186, 210, 212–214; "The Excellency
and Grounds of the Corpuscular or
Mechanical Philosophy," 204; *The
Sceptical Chymist*, 161, 167, 170–174,
187, 201–203, 208;
Brass, 187, 196–197; sphere, 199–200
British empiricism, 224
Burden of proof, 189, 209
Butterfield, Herbert, 7–8, 11, 218
Butters, Suzanne, 18

Calcination, 32–33, 41–42, 44, 78, 89–90, 129, 223
Camphor, 213–214
Cardano, Girolamo, 100
Carneades (Boylean interlocutor), 171, 202
Caterpillars, 146, 149
Catholic affections. *See* mechanical affections
Cementation, 34
Chalcanthum, 70
Chalmers, Alan, 179, 181, 200
Chemical bond, xi–xii, 27, 29, 134
Child, Robert, 15
Chrysophthoroi, 64
Chrysopoeia, xi, 11, 65, 68, 104–105, 144–145. *See also* transmutation
Chrysulca. See nitric acid
Cinnabar, 173
Clericuzio, Antonio, 3–4, 178–185
Clocks, 187, 205
Cloning, 16
Coburg, 68
Coimbrans, 116, 118–119
Cole, Michael, 18
Color, 80–81, 124–125, 153, 181–185, 188, 195, 202, 206
Combustion, 220; as *diakrisis*, 131–132; of green wood, 119–120
Complexion, 95
Composites, are dissolved into the things out of which they are made, 49
Composition, 28, 44, 46, 55–57, 59–60, 62, 64, 120; hierarchical, 98, 127, 172–173, 179–180, 189; very strong (see *aggregate corpuscles; fortissima compositio*)
Compound, xi
Compounds, 4, 27. See also *fortissima compositio*
Concretio, 124, 130
Copernicus, 7
Copper, 32–33, 89, 121, 173–174
Crama, 102–103
Creation, 114–115

Crucible, 205
Cupellation, 34, 198

Dalton, John, 5, 46, 97, 221–222
Daniel, Dane, 18
Darnel, 145, 149
Dator formarum, 95
De Witt, Lloyd, 8
Dear, Peter, 10–13
Debus, Allen G., 17
Decompounded matter, 179
Dee, John, 11, 74; *monas hieroglyphica*, 11, 74
Degeneration, 145
Democritus: and Democriteans, xi, 20, 24, 69–70, 73–77, 79, 81, 85, 87, 91–93, 104, 106, 125, 129, 132, 144–145, 158, 163–164, 218; in agreement with Aristotle, 75–78, 80–81, 92–94; misrepresented as a point atomist by Aristotle, 162, 164; *philosophia symbolica* of, 74, 77; *Physika kai mystika*, 70, 74
Demonology, 16–17
Descartes, René, 6–7, 13, 19, 131, 152, 157, 160, 162–163, 178–179, 223
Design, principle of, 147
Diakrisis. See synkrisis
Digby, Kenelm, 160
Digestion, 130, 132
Dijksterhuis, E. J., 7, 175
Disguises, assumed by chemicals, 203
Distillation, 34, 44, 72, 78, 89, 211, 223
Dogs, 170
Döring, Michael, 133
Duclo, Gaston, 105

Echeneis, 139
Eduction. *See* form
Efferarius, Frater, 44
Effluvia, 144, 169–170
Eggs, 49, 51, 61, 146–147
Electrum, 99, 120, 166–168, 193–194
Elements, four, 5, 27, 36–43, 50–52, 54, 56, 58, 62, 71–72, 75–77, 95–96, 98,

106–107, 112–115, 117–118, 128, 138–139, 171–172; are not responsible for most qualities, 96; cannot act beyond the capacity of their form, 140; have forms that can be remitted, 110–111, 118–119, 148; Lavoisian elements, 221–222; remain intact in a mixture, 95; remain *virtualiter* in a mixture, 37, 107, 113, 117. *See also* primary qualities; resolution

Eleutherius (Boylean interlocutor), 202

Emmer wheat, 149–150

Empedocles, 28, 76

Empiricism, 136–139, 221, 223

Engines. *See* machines

Epicurus, 25, 101, 141, 158, 171–173

Erastus, Thomas, 19, 45–65, 66, 104, 114, 120, 212

Essence, 57, 93, 129, 149, 163, 166, 218; in Boyle's work, 195–209; as convention of qualities, 200

"Essential structure." *See* essence

Exhalation, 131–132

Experiment and Aristotelianism, 96

External appearance, as opposed to "internal form," 124–125, 130

"Extra-essential" properties, 197–208

Fernel, Jean, 140, 143, 147

Ficino, Marsilio, 77, 140

Figure, 171

Filberts, 204

Filtration, 89, 100, 121, 123, 151, 166–167, 173–174, 193, 201

Final causes. *See* teleology

Fire, 202–203; as mechanical agent, 203; as violent agent, 47, 80. *See also* analysis

Flesh, 163

Flint, 196

Flour, 205–206

Fludd, Robert, 191

Fluidity, 204–205

Form, 5, 13, 55–56, 64, 99, 102–104, 136, 195, 213, 220; degrees of,

109–110; "dictatorship" of substantial form, 225; divine character of, 123–124, 137, 147–148; educed from matter, 147; as efficient cause, 144, 146, 153, 208; "external" form, 133, 145; form of the mixture, 37, 53, 58, 60, 107–124, 145; *forma cadaveris*, 42–43, 109–110; *forma completiva*, 54; *formae fractae* or *refractae* ("broken" forms), 110–111, 115, 118–119; as "hands of God," 123–124, 147, 152, 218; immutability of substantial forms, 148, 159; "internal" form, 122, 124, 130, 145, 149; only acts by means of qualities, 142; source of qualities in general, 129; substantial form, 35–43, 54, 57, 60, 64, 81, 107–125, 137–153, 159, 168, 183, 186, 200, 208–212; substantial form descends from the heavens, 95; unknowability of substantial forms, 137, 139, 186

Fortissima compositio, xi, 27, 29, 31, 43–44, 60, 80, 121

Fracastoro, Girolamo, 140, 143

Franklin, Benjamin, 15

Galen, 63–64, 139, 141

Galileo, 7, 158

Gassendi, Pierre, 6, 13–14, 157, 160, 162, 191–194, 223; *Animadversiones in decimum librum Diogenis Laertii*, 192

Geber: and Geberians, xi, 11, 13–16, 19, 26–35, 43, 45–46, 60, 65, 69, 80–81, 85, 89–90, 96, 97–98, 104, 106, 114, 127–129, 132, 144, 152, 165–166, 168, 173, 190, 220; 223–224; idol and god of the alchemists, 48, 64, 66; *Summa perfectionis*, 13–14, 26–35, 43, 60, 66, 69, 80, 90, 98, 223

Generation and corruption, 37–44, 99, 107, 118, 146, 171; circular, 117; of man, 71; versus mixture, 53, 59, 107. *See also* spontaneous generation

Genus, 61

Glass, 202–203, 207, 211

Glauber, Johann Rudolph, 210

Goethe, Johann Wolfgang von, 17
Gold, 32, 41, 55, 58, 60, 88, 99, 103,
    112, 120, 145, 150, 166–168,
    173–174, 193–194, 201
Goslar, 89
Guibert, Nicolas, 69, 104
Gunpowder, 135–136, 138, 188,
    191
Gymnasium Casimirianum, 68, 76

Habit. *See* form
Hall, A. Rupert, 7–8, 11, 218
Hall, Marie Boas, 6, 8, 10–11, 157–158,
    218, 223
Halleux, Robert, 18
Halonitrum, 70
Hammer-Jensen, Ingeborg, 67
Hartlib, Samuel, 14, 159
Harvard College, 15
Harvey, William, 46
Heat, 203, 206; Bacon's theory of, 203;
    internal motion of parts, 197
Helmont, Joan Baptista Van, 15–16, 64,
    181, 211
Hermetic tradition, 12
Hermetici philosophi, 76–78, 81
Hippocrates, *Epistola ad Damagetum*,
    73–74
Homoeomerity, 30–31, 163–166
Homogeneity, xii, 27, 30–31, 43, 47, 56,
    63–64, 80, 85, 100, 107, 131, 152,
    165–166, 169. *See also* homoeomerity;
    uniformis substantia
Homunculus, 17
Horn, 188
Humors, 141–142
Hunter, Michael, 159
Hydrochloric acid, 98, 174
Hylomorphism, 13, 36, 78, 129, 137,
    148, 151–152, 161, 208, 212, 218,
    220, 224–225
Hylozoism, 85

Ice, 206
Identity: *in numero*, 55, 58, 118, 167,
    173, 201; *in specie*, 118

Immutation, 130–132, 146, 148, 182,
    190
Insects, 145, 149
Iron, 32–33, 89, 187, 195–197
Ivy-wood, 102, 116

Jābir ibn Hayyān, 26
Jaeger, Werner, 67
Jena, 68
Jessenius, Johann, 139–140
Jesuits, 111, 224
John XXI (pope), 26

Kahn, Didier, 17
Kaufmann, Thomas DaCosta, 18
Kuhn, Thomas, 159

Lasswitz, Kurd, 85
Lavoisier, Antoine Laurent, 16, 46, 97,
    219–222
Lead, 33, 60, 62, 104, 133, 173–174,
    187; acetate (*see* sugar of lead); color
    changes in, 202
Lemery, Nicolas, 129, 131
Leucippus, xi, 24, 69
Libavius, Andreas, 10–11, 19, 68–81,
    85, 87, 89, 92–94, 105, 218; *Alchymia*,
    69, 71, 74–75, 89–90; *Alchymia
    triumphans*, 75–76; *De mundi
    corporumque mixtorum elementis*,
    76–77; *Rerum chymicarum epistolica
    forma*, 71, 73
Light, 5, 169, 184, 188, 196, 206–207
Like-to-like, 32, 132–133, 135
Limestone, 129
Locke, John, 15, 224
Lodestone, 139
Lucretius, 25, 31, 141
Lull, Ramon, 11, 43
Lutheransm, 86
Lüthy, Christoph, 73
Lynceus, 28

Machines, 185–188, 205, 209;
    mechanical advantage, 186–187
Magistery, 131

Magnenus, Johannes Chrysotomos, 160
Maier, Anneliese, 53, 107
Mandelbaum, Maurice, 204
Manifest qualities. *See* primary qualities
Marble, 196
Materia proxima, 52, 60
McMullin, Ernan, 176–177
Mechanical affections (size, shape, and motion), 180–182, 184–187, 189, 200
Mechanical Philosophy, 3, 5–6, 12, 19, 23–24, 125, 151, 157–215, 218, 222–225
Mechanism, 67, 134, 161, 175–179, 190, 215, 222, 225; criteria of a mechanical explanation, 175–180; mechanical as opposed to chymical explanation, 182–185
Menstruum, 133, 164, 166–167, 173, 201
Mercurialists, 14
Mercuric chloride, 182, 192
Mercury alone theory, 14, 34, 60
Mercury oxide, 219
Mercury, 14, 29–33, 38–39, 58, 60, 65, 80, 98, 123, 128, 131, 135, 145, 174, 183, 196; mercury sublimate (*see* mercuric chloride)
Mersenne, Marin, 158
Metakinēsis, 131
Microscope, 169, 171
Milk, 75, 163, 165–166, 168, 211
Mills, 205
Minima, 27, 30–31, 90, 98, 101, 127, 130, 132, 136, 172, 191; *minima inexistentia*, 78; *minimae atomi*, 122; of their own genus, 93, 128, 172, 197, 203, 207, 214
Mites, 169, as engines, 169
Mixis. See mixture
Mixture, xii–xiii; 4–5, 27, 37, 43–46, 48, 53, 59–60, 63–65, 77–78, 85, 100, 107–123, 127, 132, 136, 138, 141, 144–145, 147, 150, 152–153, 163–166, 183, 212–213; imperfect, 102, 119–121; ingredients cannot be

retrieved from, 4–5, 45–46; *mixtio ad sensum*, 121; *mixtio per minima*, 27, 30–31, 136, 165, 167–168; perfect, xiii, 121; requires resolution up to the four elements or the prime matter, 37, 106, 224; Scaligerian concept of, 101–102
Modus, 124, 130–131
Moleculae, 192
Molecule, 28
Moran, Bruce, 3, 18
Moses, 171–172, 174
Motion, internal, 196–197, 205–206; local, 153, 171, 178, 188; transmission of, 176–178; vibratory, 197; wave, 196. *See also* heat; sound
Mulberry, 146
Must, 72

Nature, 42–43, 47–48, 51–52, 62–64, 88, 96, 104, 114–115, 127, 139, 147, 149; alchemists wage war on, 64; as *ordinaria Dei potestas*, 63, 152; theater of, 145
Negative-empirical principle, 96, 99, 127, 164, 223. *See also* analysis; atom
Neoplatonism, 77, 140
Nescience, 137, 209
Newton, Isaac, 4–5, 7, 15–16, 44, 134–135, 221, 224–225
Niter, 70, 135, 138, 188, 190, 205–206, 210–212; fixed, 210; spirit of, 210
Nitric acid, 64, 98, 103, 112–115, 121–123, 127, 134, 150, 153, 166–167, 173–174, 193–194, 210–211
Numerical identity. *See* identity
Nummedal, Tara, 18

Oats, 149
Occult qualities, 139–140, 142, 144, 148, 170
Odors, 214
Oil, 135–136
Oil of tartar. *See* salt of tartar
Oil of vitriol. *See* sulfuric acid

Onkoi, 66–67
Osler, Margaret, 10

Pantheo, Giovanni Agostino, 44
Paracelsus, 17, 45, 47, 67–68; *tria prima* of, 47–48, 94, 96, 127, 157, 160, 187
Parmenides, 24, 55, 57
Paul of Taranto, 26, 35–44, 100, 108, 218
Pereira, Benedictus, 88
Perfume, 169
Peripatetics. *See* Aristotle, and Aristotelians
Philalethes, Eirenaeus, 14, 44, 222. *See also* Starkey, George
Philoponos, John, 52, 116–117
Philosophers' stone, 34, 60
Phthisis, 142
Pitch, burning, 132
Plague, 140, 142–143
Plato, 87
Pliny, 102, 116
Pluralists, 36–43, 54–58, 64, 108–111, 113–115
Plurality of forms theory. *See* pluralists
Pneuma, 170
Poisons, 137, 142–143
Poroi, 66–67
Potency. *See* act and potency
Precipitation, 89, 103, 131, 134–135, 166–167, 194, 223; red precipitate (*see* mercury oxide)
President's Council on Bioethics, 16
Priestly, Joseph, 219–220
Prima mista. See *prima mixta*
Prima mixta, 127, 172, 176, 179–180
Prima naturalia, 172, 176, 179–182, 185, 188–189, 200–201, 203; 209, 220, 223
Primary clusters. See *prima mixta*
Primary concretions. See *prima mixta*
Primary qualities, scholastic, 36–37, 56, 78, 111–112, 138, 142, 187; inadequate for explaining many phenomena, 143; remitted in a mixture, 95. *See also* elements

Prime matter, 36–38, 58, 75, 106
Principe, Lawrence M., xi, 3, 10, 14–15, 18, 159, 221
Principle of parsimony, 189, 209–210
Privation, 50–54, 60, 212
Pugna elementorum, 138
Putrefaction, 131, 143, 146
Pyle, Andrew, 176–178
Pyrophilus (Boylean interlocutor), 183

Qualitates fractae. See Avicenna
Qualities, "chymical," 200; "flow" from substantial form, 129, 136, 139, 144, 153, 183, 187, 211; "material," 141; particular, 175; "spiritual," 141; of the whole substance, 139, 141, 143. *See also* essence; occult qualities; primary qualities
Quicklime, 138
Quiddity, 57

Radishes, 149
Rational soul, 54
Redintegration: of amber, 212; of niter, 210–212; of turpentine, 212; of vitriol, 212
Reditus principle, 50–54, 60, 64, 104–105, 115–116, 121, 124, 212
Reduction to the pristine state, 24, 35, 41–43, 98–100, 103–105, 112–115, 123–124, 149–151, 158–159, 161, 167–168, 190–215, 217–219, 223
Reduction, xiii, 105, 150–152; of camphor, 213, 223
Reductionism, 125, 129, 151, 176–177, 181, 225
Reflection, 184, 206
Regulus martis, 173
Resolution, 27, 29, 117, 171; up to four elements or prime matter, 37, 58, 60, 106, 109–111
Resurrection, 104, 115
Reversible reactions, xii–xiii, 113, 121, 159, 161, 220, 224
Rhodiginus, Caelius, 87

Riolan, Jean, 70
River-lettuce, 118
Rocke, Alan, 221–222
Rosicrucians, 12
Rosin, 207
Ross, David, 67
Rossi, Paolo, 11–12
Rotenburg, 68
Roughness, 196
Roundness, 199–200
Royal College of Physicians, 15
Royal Society, 34
Rudolph, Hartmut, 18

Sala, Angelus, 105, 135
Salerno, School of, 31
Saline spirits, 132–134
Salt of tartar, 112–114, 134–135, 144, 166–167, 182–184, 192
Salt, 72, 128, 132–133, 174, 191, 205; acid, 183–184, 198; alkaline, 184, 198; of blood, 165, 195; urinous, 198; vinous, 133
Sand, 205–206
Sargent, Rose-Mary, 10, 159, 178
Scaliger, Julius Caesar, 99–104, 124, 143; on unknowability of substantial forms, 137. *See also* mixture
Schaffer, Simon, 8
Scheidung, 45, 68
Scientific Revolution, 2–3, 7–13–14, 18, 26, 217–218, 223–225
Scotus, John Duns, 107–108
Semina, 76–78, 81, 85, 140, 143, 146–147, 151; and ancient atomism, 141; of disease, 140–144
Seminal principles, 77, 81, 147
Seminaria. See semina
Seminia. See semina
Senna, 192–193
Sennert, Daniel, xi, 1–2, 8, 14, 20, 23–25, 27, 42, 67, 71, 74, 79–80, 85–153, 158, 160–170, 174–176, 189–191, 193–194, 197, 203,
207–215, 217–225; claim that "all things are composed of atoms," 91, 106; works of: *De chymicorum cum Aristotelicis et Galenicis consensu ac dissensu* (1619), 23, 86, 91–96, 99, 103, 106, 111–112, 122–124, 127, 130–131, 144–150, 162, 211; *De chymicorum cum Aristotelicis et Galenicis consensu ac dissensu* (1629), 92, 94–95, 100, 130, 132–136; *De differentiis morborum disputatio*, 142; *De febrium malignarum natura & causis disputatio*, 142; *De methodo medendi disputatio*, 141–142; *De pestilentia disputatio*, 142–143; *Epitome naturalis scientiae* (1600), 87–89, 100, 109–111, 140, 147; *Epitome naturalis scientiae* (1618), 89; *Hypomnemata physica*, 20, 93, 100, 122, 127–128, 134, 148–151, 162, 165, 172; *Institutiones medicinae*, 89, 92, 129–130, 132
Sensitive soul, 54, 56
Severinus, Petrus, 77, 80, 141
Shackelford, Jole, 18
Shapin, Steven, 2, 8–12, 157–158, 224
Shininess, 197
Silkworms, 146
Silver, 60, 62, 99, 103, 112–115, 120–123, 128, 145, 150–151, 153, 166–168, 173, 187, 195–197, 211, 214; not transmuted into gold, 193–194
Smith, Pamela, 3, 18
Smolnitz, 89
Smoothness, 196
Snake, 149
Sound, 196
Spagyria, 46, 65, 67–68, 79, 171; from *span* and *ageirein*, 69
Species, 55–56, 58, 60, 62, 144–147, 149–150, 152, 198
Specific identity. *See* identity
Spirit of niter. *See* nitric acid
Sponges, 118

Spontaneous generation, 146–147, 151

Springiness, 195, 209

Stagirite. *See* Aristotle

Stahl, Georg Ernst, 222

Starkey, George, 9, 15–16, 159, 222

Steel, 187, 196

Stengers, Isabelle, 96

Stoics, 74, 170

Stone, 187, 197

Strings, gut, 196

Structural explanation, 125, 136, 138, 148, 151, 176–177, 220, 223–224

Structure, as opposed to form, 124, 129, 134, 138, 144–145, 149

Suavius, Leo, 69

Sublimation, xiii, 29, 34, 44, 78, 90, 127–129, 223

Subordination of forms theory, 145–151

Substance, in scholastic philosophy, 35–44, 113, 148, 199; substantial change, 203

Sugar of lead, 133

Sulfur, 14, 29–33, 38–39, 58, 80, 98, 128, 135, 173, 188, 191; flowers of, 128

Sulfuric acid, 98, 144, 173, 182, 184, 192, 213

Sympathy, 140, 144

Syndiacritical theory, 67

Synkrisis and *diakrisis*, 69–71, 73, 78–79, 85, 89–93, 104, 106, 125, 129–134, 144–145, 148, 153, 182, 190, 211, 214–215, 218

Synthesis. See composition

Syphilis, 140–141

Taxis, 131

Teleology, 67, 177

Telle, Joachim, 17

Temperamentum, 56

Tests, for chymical species, 198; indicator, 198

Texture, 171, 180, 182–185–187, 191–192, 194–198, 202, 206, 207, 212, 214

Thesis, 131

Thorndike, Lynn, 34

Three principles. *See* Paracelsus

Tin, 33, 121, 173–174

Toletus, Franciscus, 88, 116–118, 121

Torpedo, 139

Transdiction, 204–208; by analogy, 204–205; by substantial identity, 205–208

Transmutation, 38, 73, 104, 115–116, 148, 212, 220; of iron into copper, 89. *See also* Chrysopoeia

Transposition, 176, 188–189, 203

Turnips, 149

Turpentine, 212

Uniform catholic matter, 163, 171, 200

Uniformis substantia. See Homogeneity

Union, 27, 32, 100–101, 103, 133, 167–168

Unitists, 35–43, 55–58, 108, 113–115, 218, 224

Unity of forms theory. See unitists

Vacuum, 171

Vegetative soul, 40–43, 56

Vinegar, 133

Vitrification, 131

Vitriol, 212; springs of, 89

Walnuts, 204

Watch, 209

Webster, Charles, 159

Weeks, Andrew, 18

Westfall, Richard, 175

Wheat, 145, 149–150, 205–206

White, William, 15

Whiteness, 184, 188, 195, 199, 206–207

Wine,72, 163, 166, 168; cannot return from vinegar, 50–51, 60–61, 115; mixed with water, 52, 102, 115–119; spirit of, 133, 165; transmuted into water, 193

Winthrop, John Jr., 15

Wittenberg, University of, 81, 85–86, 91, 100, 106, 139, 148, 160, 166

Wood, 40–43, 56, 187, 197

Worsley, Benjamin, 210

Yates, Frances, 11–12

Zabarella, Jacobus, 88, 108–109–111, 118–119, 145, 148; *De rebus naturalibus*, 109